The Curve Shortening Problem

The Curve Shortening Problem

Kai-Seng Chou
Xi-Ping Zhu

CRC Press
Taylor & Francis Group
Boca Raton London New York

CRC Press is an imprint of the
Taylor & Francis Group, an **informa** business
A CHAPMAN & HALL BOOK

CRC Press
Taylor & Francis Group
6000 Broken Sound Parkway NW, Suite 300
Boca Raton, FL 33487-2742

First issued in paperback 2019

© 2001 by Taylor & Francis Group, LLC
CRC Press is an imprint of Taylor & Francis Group, an Informa business

No claim to original U.S. Government works

ISBN-13: 978-1-58488-213-8 (hbk)
ISBN-13: 978-0-367-39753-1 (pbk)

Library of Congress Cataloging-in-Publication Data

Chou, Kai Seng.
    The curve shortening problem / Kai-Seng Chou, Xi-Ping Zhu.
      p. cm.
    Includes bibliographical references and index.
    ISBN 1-58488-213-1 (alk. paper)
    1. Curves on surfaces. 2. Flows (Differentiable dynamical systems) 3. Hamiltonian
sytems. I. Zhu, Xi-Ping. II. Title.

QA643 .C48 2000
516.3′52—dc21

00-048547

Library of Congress Card Number 00-048547

Visit the Taylor & Francis Web site at
http://www.taylorandfrancis.com

and the CRC Press Web site at
http://www.crcpress.com

CONTENTS

vi

PREFACE

A geometric evolution equation (for plane curves) is of the form

$$\frac{\partial \gamma}{\partial t} = f \boldsymbol{n} \ , \qquad (*)$$

where $\gamma(\cdot, t)$ is a family of curves with a choice of continuous unit normal vector $\boldsymbol{n}(\cdot, t)$ and f is a function depending on the curvature of $\gamma(\cdot, t)$ with respect to $\boldsymbol{n}(\cdot, t)$. Any solution of $(*)$ is invariant under the Euclidean motion. The simplest geometric evolution equation is the eikonal equation when f is taken to be a non-zero constant. The next one is the curvature-eikonal flow when f is linear in the curvature. It includes the curve shortening flow (CSF)

$$\frac{\partial \gamma}{\partial t} = k \boldsymbol{n} \ ,$$

as a special case. Let $L(t)$ be the perimeter of a family of closed curves $\gamma(\cdot, t)$ driven by $(*)$. We have the first variation formula

$$\frac{dL}{dt}(t) = - \int_{\gamma(\cdot, t)} f k \, ds \ .$$

Therefore, the CSF is the negative L^2-gradient flow of the length. When $\gamma(\cdot, t)$ is also embedded, its enclosed area satisfies

$$\frac{dA}{dt}(t) = -2\pi \ .$$

Thus, any embedded closed curve shrinks under the flow and ceases to exist beyond $A(0)/2\pi$. The following two results completely characterize the motion.

Theorem A (Gage-Hamilton) The CSF preserves convexity and shrinks any closed convex curve to a point. Furthermore, if we dilate the flow so that its enclosed area is always equal to π, the normalized flow converges to a unit circle.

Theorem B (Grayson) The CSF starting at any closed embedded curve becomes convex at some time before $A(0)/2\pi$.

From the analytic point of view, the curvature of $\gamma(\cdot, t)$ satisfies

$$k_t = k_{ss} + k^3 ,$$

where $s = s(t)$ is the arc-length parameter of $\gamma(\cdot, t)$. This is a non-linear heat equation with superlinear growth. It is clear that the curvature must blow up in finite time. However, it is the geometric nature of the flow that enables one to obtain precise results like these two theorems. On the other hand, the CSF is a special case of the mean curvature flow for hypersurfaces. It turns out that, although Theorem A continues to hold for the mean curvature flow, Theorem B does not. This makes planar flows special among curvature flows.

After Theorems A and B, subsequent works on the CSF go in two directions. One is to study the structure of the singularities of the flow for immersed curves, and the other is to consider more general planar flows. In this book, we present a complete treatment on Theorem A and Theorem B as well as provide some of generalizations. There are eight chapters. We outline the content of each chapter as follows: In Chapter 1, we discuss basic results such as local existence, separation principle, and finiteness of nodes for the general flow (∗) under the parabolic assumption. In Chapter 2, we describe special solutions of the CSF which arise from its Euclidean and scaling invariance: travelling waves, spirals, and contracting and expanding self-similar solutions. These solutions will become important in the classification of singularities for the CSF. Theorem A is proved in Chapter 3. In the same chapter, we also study the anisotropic curvature-eikonal flow

$$\frac{\partial \gamma}{\partial t} = (\Phi(\boldsymbol{n})k + \Psi(\boldsymbol{n})) , \quad \Phi > 0 .$$

This flow may be viewed as the CSF in a Minkowski geometry when $\Psi \equiv 0$ and its general form is proposed as a model in phase transition. Depending on the inhomogeneous term Ψ, it shrinks to a point, expands to infinity, or converges to a stationary solution. We determine its asymptotic behaviour in all these cases. In Chapter 4, we study the anisotropic generalized CSF

$$\frac{\partial \gamma}{\partial t} = \Phi(\boldsymbol{n})|k|^{\sigma-1}k\boldsymbol{n} , \quad \Phi > 0 , \ \sigma > 0 ,$$

following the work of Andrews [8] and [10]. Analogues of Theorem A are proved for $\sigma \in [1/3, 1)$. When $\sigma = 1/3$ and $\Phi \equiv 1$, the flow is affine invariant and is proposed in connection with image processing and computer vision. Beginning from Chapter 5, we turn to non-convex curves. First, we present a relatively short proof of Theorem B which is based on the blow-up and the classification of singularities. This approach has been successfully adopted in many geometric problems including nonlinear heat equations, harmonic heat flows, Ricci flows, and the mean curvature flow. Next, we present Grayson's geometric approach where the Sturm oscillation theorem is used in an essential way in Chapter 6. Though strictly two-dimensional, it is powerful and works for a large class of uniformly parabolic flows $(*)$. In Chapter 7, we discuss how the CSF can be used to prove the existence of embedded, closed geodesics on a surface. Finally, in Chapter 8, we study the isotropic generalized CSF and establish an almost convexity theorem when $\sigma \in (0, 1)$. Whether the convexity theorem holds for this class of flows remains an unsolved problem.

Many interesting results on $(*)$ have been obtained in the past fifteen years. It is impossible to include all of them in a book of this size. Apart from a thorough discussion on Theorem A and B, the choice of the rest of the material in this book is rather subjective. Some are based on our work on this topic. To balance things the we sketch the physical background, describe related results, and occasionally point out some unsolved problems in the notes which can be found at the end of each chapter. We hope that the reader can gain a panoramic view through them. We shall not discuss the level-set approach to curvature flows in spite of its popularity. Here we are mainly concerned with singularities and asymptotic behaviour of planar flows where the classical approach is sufficient.

Thanks are due to Dr. Sunil Nair for proposing the project, and to Ms.Judith Kamin for her effort in editing the book. We are also indebted to the Earmarked Grant of Research, Hong Kong, the Foundation of Outstanding Young Scholars, and the National Science Foundation of China for their support in our work on curvature flows, some of which has been incorporated in this book.

Chapter 1

Basic Results

In this chapter, we first establish the existence of a maximal solution and some basic qualitative behaviour such as the separation principle and finiteness of nodes for the general flow (1.2). These properties are direct consequences of the parabolic nature of the flow. For the reader's convenience, we collect fundamental results on parabolic equations in Section 2. In particular, the "Sturm oscillation theorem," which is not found in standard texts on this subject, will play an important role in the removal of singularities of the flow. In Section 3, we derive evolution equations for various geometric quantities of the flow. They will become important when we study the long time behaviour of the flow.

1.1 Short time existence

We begin by recalling the definition of a curve. An immersed, C^1-curve is a continuously differentiable map γ from I to $\mathbb{R}^2$ with a non-zero tangent $\gamma_p = d\gamma/dp$. Throughout this book, I is either an interval or an arc of the unit circle S^1, and a curve always means an immersed, C^1-curve unless specified otherwise. The curve is closed if I is the unit circle. It is embedded if it is one-to-one. Given a curve $\gamma = (\gamma^1, \gamma^2)$, its unit tangent is given by $\boldsymbol{t} = \gamma_p/|\gamma_p|$ and its unit normal, $\boldsymbol{n}$, is given by $(-\gamma_p^2, \gamma_p^1)/|\gamma_p|$. When γ is an embedded closed curve and $\boldsymbol{t}$ runs in the counterclockwise direction, $\boldsymbol{n}$ is the inner unit normal. The **tangent angle** of the curve is the angle θ between the unit tangent and the positive x-axis. It is defined as modulo 2π. However, once the tangent angle at a certain point on the curve is specified, a choice of continuous tangent angles along the flow is determined uniquely. The curvature of γ with respect to $\boldsymbol{n}$, k, is defined via the Frenet formulas,

$$\frac{d\boldsymbol{t}}{ds} = k\boldsymbol{n}, \qquad \frac{d\boldsymbol{n}}{ds} = -k\boldsymbol{t}, \tag{1.1}$$

where $ds = |\gamma_p| dp$ is the arc-length element. Explicitly we have

$$k = \frac{\gamma_{pp}^2 \gamma_p^1 - \gamma_{pp}^1 \gamma_p^2}{|\gamma_p|^3}.$$

We shall study the flow

$$\frac{\partial \gamma}{\partial t} = F(\gamma, \theta, k)\boldsymbol{n}, \quad (p, t) \in I \times (0, T), \quad T > 0, \tag{1.2}$$

where $F = F(x, y, \theta, q)$ is a given function in $\mathbb{R}^2 \times \mathbb{R} \times \mathbb{R}$, 2π-periodic in θ. A **(classical) solution** to (1.2) is a map γ from $I \times (0, T)$ to $\mathbb{R}^2$ satisfying (i) it is continuously differentiable in t and twice continuously differentiable in p, (ii) for each t, $p \longmapsto \gamma(p, t)$ is a curve, and (iii) γ satisfies (1.2) where $\boldsymbol{n}$ and k are respectively the unit normal and curvature of $\gamma(\cdot, t)$ with respect to $\boldsymbol{n}$. Given a curve γ_0, we are mainly concerned with the following Cauchy problem:

To find a solution of (1.2) which approaches γ_0 as $t \downarrow 0$.

For simplicity, let's assume F is smooth in all its arguments. It is called **parabolic** in a set E if $\partial F/\partial q(x,y,\theta,q)$ is positive for all (x,y,θ,q) in E and **uniformly parabolic** in E if there are two positive numbers λ and Λ such that

$$\lambda \leqslant \frac{\partial F}{\partial q} \leqslant \Lambda \, ,$$

holds everywhere in E. Further, F is called **symmetric** in E if

$$F(x,y,\theta+\pi,-q) = -F(x,y,\theta,q) \, ,$$

holds in E. When the set E is not mentioned, it is understood that F is parabolic, uniformly parabolic, or symmetric in its domain of definition. Among the flows described in the preface, the curve shortening flow is uniformly parabolic and symmetric. The generalized curve shortening flow is parabolic, symmetric but not uniformly parabolic in $\mathbb{R}^3 \times \mathbb{R}\backslash\{0\}$ when σ is not equal to 1. The anisotropic curvature-eikonal flow is uniformly parabolic but not symmetric in general. This is clear from its physical meaning: Reversing the orientation of the curve means interchanging the phases. Although the initial phase boundaries are identical and the difference in temperature is the same, the flows evolve in different ways.

Recall that a reparametrization of a curve γ is another curve $\gamma'(p) = \gamma(\phi(p))$ where ϕ is a diffeomorphism. The reparametrization is orientation preserving if ϕ' is positive and orientation reversing if ϕ' is negative. It is an important fact that (1.2) is invariant under any orientation preserving parametrization. In fact, an orientation preserving reparametrization $\gamma'(p,t) = \gamma(\phi(p'),t)$ solves (1.2) if γ itself is a solution. Equation (1.2) is also invariant under any orientation reserving reparametrization when F is symmetric. In particular, for an embedded solution, that is, $\gamma(\cdot,t)$ is embedded for each t, the flow is independent of which parametrization is used in the beginning, and so it only depends on the geometry of the initial curve. In this sense, (1.2) is geometric. However, this may no longer be valid for immersed curves. In an example following the proof of Proposition 1.6, one will see an example which really depends on the parametrization of the initial curve.

What is the type of (1.2)? At first sight, the most natural way is to view it as a system of two equations for the two unknowns γ^1 and γ^2. To examine its type, we need to linearize the system at a solution γ. Thus, let $\gamma(\varepsilon)$ be a family of solutions of (1.2) satisfying $\gamma(0) = \gamma$ where γ is parameterized in arc-length. Then $\Gamma = d\gamma/d\varepsilon(0)$ satisfies

$$\frac{\partial \Gamma}{\partial t} = \frac{\partial F}{\partial q} M \frac{\partial^2 \Gamma}{\partial p^2} + g \, ,$$

where

$$M = \begin{bmatrix} (\gamma_p^2)^2 & -\gamma_p^1 \gamma_p^2 \\ -\gamma_0^1 \gamma_p^2 & (\gamma_p^1)^2 \end{bmatrix} ,$$

and g depends on Γ, Γ_p but not on Γ_{pp}. It is clear that M is a non-negative matrix and has a unique null direction given by γ_p, reflecting the invariance of (1.2) under reparametrization. So even if F is parabolic, (1.2) is never parabolic when viewed as a system. To retrieve parabolicity, we need to fix a parametrization and express (1.2) as a single parabolic equation. There are several ways to achieve this goal. The simplest one is to express the solution in local graphs. Suppose during some time interval each $\gamma(\cdot, t)$ is the graph of a function $u(x, t)$ defined over some interval J. If u_x is bounded, using the inverse function theorem we can write $\gamma(p, t) = (x, u(x, t))$

where $x = x(p,t)$ and u is as regular as γ. We have

$$\frac{\partial \gamma}{\partial t} = \frac{\partial x}{\partial t}(1, u_x) + (0, \frac{\partial u}{\partial t}) \ .$$

Taking the inner product with $\boldsymbol{n} = (-u_x, 1)/\sqrt{1 + u_x^2}$ (here we assume the orientation of the curve is along the positive x-axis), we see that u satisfies the equation

$$\frac{\partial u}{\partial t} = \sqrt{1 + u_x^2}\, F(x, u, \theta, k) \ , \tag{1.3}$$

where now $\tan \theta = u_x$ and k is given by

$$k = \frac{u_{xx}}{(1 + u_x^2)^{3/2}} \ .$$

Clearly, when the gradient of u is bounded, (1.3) is a parabolic equation if and only if F is parabolic.

Another useful way to obtain a single parabolic equation from (1.2) is to represent the flow as graphs over some fixed curve. When γ_0 is smooth, we may take the fixed curve to be γ_0. But, it is always more convenient to take it to be a smooth or analytic curve close to γ_0. For a fixed closed, smooth curve Γ near γ_0, we can express the solution of (1.2) starting at γ_0 as

$$\gamma(p, t) = \Gamma(p) + d(p, t)\boldsymbol{N}(p) \ ,$$

where $\boldsymbol{N}$ is the unit normal of Γ. When t is small, d is small and d_p is bounded. Let's write down the equation for d. To simplify the computation we shall assume Γ is parametrized by the arc-length, i.e., $|\Gamma_p| = 1$. First, by (1.1),

$$\boldsymbol{t} = \frac{\Gamma_p + d_p \boldsymbol{N} - \kappa d\boldsymbol{T}}{|\gamma_p|} \ ,$$

$$|\gamma_p|^2 = (1 - \kappa d)^2 + d_p^2 \ ,$$

where $\boldsymbol{T}$ and κ are respectively the unit tangent and curvature of Γ. We also have

$$k = \frac{(1-\kappa d)d_{pp} + 2\kappa d_p^2 + \kappa_p d d_p - 2\kappa^2 d + \kappa^3 d^2}{[(1-\kappa d)^2 + d_p^2]^{3/2}} . \tag{1.4}$$

It follows that d satisfies

$$d_t = \frac{1-\kappa d}{[(1-\kappa d)^2 + d_p^2]^{1/2}} F(\gamma,\theta,k) . \tag{1.5}$$

When F is parabolic, (1.5) is parabolic as long as $\kappa d < 1$ and d_p is bounded. Local solvability of the Cauchy problem for (1.5) can be readily deduced from standard parabolic theory. Before formulating an existence result, we first show the equivalence between (1.2) and (1.5). In fact, the following result holds.

Proposition 1.1 *Consider the flow*

$$\frac{\partial\gamma}{\partial t} = F(\gamma,\theta,k)\boldsymbol{n} + G(\gamma,\theta,k)\boldsymbol{t} , \tag{1.6}$$

where F and G are smooth and 2π-periodic in θ. Let γ be a solution of (1.6) in $C^\infty(S^1 \times [0,T))$. There exists $\varphi : S^1 \times [0,T) \to S^1$ satisfying $\varphi' > 0$ and $\varphi(p,0) = p$ such that $\gamma'(p,t) = \gamma(\varphi(p,t),t)$ solves (1.2).

Proof: With the notation as above, we have

$$\frac{\partial\gamma'}{\partial t}(p,t) = \frac{\partial\gamma}{\partial t}(\varphi,t) + \frac{\partial\gamma}{\partial p}(\varphi,t)\frac{\partial\varphi}{\partial t}$$

$$= F(\gamma',\theta',k')\boldsymbol{n} + G(\gamma',\theta',k')\boldsymbol{t} + \frac{\partial\varphi}{\partial t}\gamma_p.$$

So γ' solves (1.2) if

$$\frac{\partial\varphi}{\partial t} = -|\gamma_p(\varphi)|\ G(\gamma',\theta',k). \tag{1.7}$$

With γ already known, (1.7) can be regarded as an ordinary differential equation where p is a parameter. From the smooth dependence
on a parameter for solutions of ODE's, we deduce the existence of a
solution φ satisfying $\varphi(p,0) = p$ and $\varphi' > 0$. $\qquad\square$

This proposition shows that the geometry of the flow (1.6) depends only on its normal velocity F while the tangent velocity merely
alters the parametrization of the flow.

Proposition 1.2 (local existence) *Consider the Cauchy problem
for (1.2) where F is smooth and parabolic, and γ_0 belongs to $C^{2,\alpha}(S^1)$
for some $\alpha \in (0,1)$. There exists a positive $\omega \leqslant \infty$ such that (1.2)
admits a solution γ in $\tilde{C}^{2,\alpha}\left(S^1 \times [0,\omega)\right)$ satisfying $\gamma(\cdot,0) = \gamma_0$. Moreover, the followings hold:*

(i) *γ is smooth (and analytic if F is analytic) in $S^1 \times (0,\omega)$,*

(ii) *if ω is finite, then $k(\cdot,t)$, the curvature of $\gamma(\cdot,t)$, becomes unbounded as $t \uparrow \omega$, and*

(iii) *if γ_0 depends smoothly (or analytically and F is analytic) on a
parameter, so does γ.*

From now on, we call the solution defined in $(0,\omega)$ the **maximal
solution** of (1.2) starting at γ_0.

Proof: Applying Fact 3 in the next section to (1.5) and then using
Proposition 1.1, we know (1.2) has a unique solution γ in $C^\infty(S^1 \times
[0,T])$, $T > 0$, $\gamma(\cdot,0) = \gamma_0$, where T depends on the $C^{2,\alpha}$-norm of
γ_0. Moreover, (i) and (iii) hold. Also, the solution can be extended
as long as the $C^{2,\beta}$-norm of γ is under control for some $\beta \in (0,1)$.

We shall show that a uniform curvature bound yields a bound on some $C^{2,\beta}$-norm for γ. In fact, we can always use a rigid motion to bring a point $\gamma(p_0, t_0)$ to the origin so that $\boldsymbol{n}(p_0, t_0) = (0, 1)$. The curvature bound ensures that there exist $\delta > 0$ and a rectangle $R = (-a, a) \times (-b, b)$ such that, for all $t \in (0, \omega)$, $|t - t_0| \leqslant \delta$, $\gamma(\cdot, t) \bigcap R$ is the graph of a function $u(x, t)$ whose first and second derivatives in x are bounded in $(-a, a)$. By differentiating (1.3), we see that the function $w = u_x$ satisfies a uniformly parabolic equation of the form

$$ w_t = \left(a(x, t) w_x \right)_x + b(x, t) w_x + c(x, t) w + f(x, t) \, , $$

where the coefficients and f are uniformly bounded. By Theorem 11.1 in Chapter 3 of Ladyzhenskaja-Solonnikov-Uralćeva [86], we conclude that there exists some $\beta \in (0, 1)$ such that $\|u\|_{\tilde{C}^{2,\beta}}$ is uniformly bounded in any compact subset of $R \times (t_0, t_0 + \delta)$. By parabolic regularity theory, all higher derivatives of u are bounded in any compact subset of $R \times (t_0, t_0 + \delta)$ as well. If the curvature of $\gamma(\cdot, t)$ is bounded near ω, we may solve (1.2) using $\gamma(\cdot, t_0)$ where t_0 is close to ω as initial curve to obtain a solution which extends beyond ω. This is impossible. Hence, the curvature must become unbounded as $t \uparrow \omega$. $\square$

Remark 1.3 The above proof has established the following estimate: *Let γ be a solution of (1.2) in $S^1 \times [0, T)$. Then, for any $\delta > 0$ and $k \geqslant 1$, there exists a constant C depending on the curvature such that*

$$ \max \left\{ \left| \frac{\partial^{i+j} \gamma}{\partial s^i \partial t^j} \right| \; : \; i + 2j \leqslant k \right\} \leqslant C \, , $$

on $S^1 \times [0, T)$, where $s = s(t)$ the arc-length parameter of $\gamma(\cdot, t)$.

To see this, let's first observe that it follows from the above proof that $\partial k / \partial s$ are bounded by k in $[0, T)$. Next, by differentiating the relations $\left|\gamma_s\right|^2 = 1$ and $\gamma_s^1\gamma_{ss}^2 - \gamma_s^2\gamma_{ss}^1 = k$ repeatedly we see that $\partial^n\gamma/\partial s^n$ can be estimated by k and its derivatives up to $(n-1)$-th order. Finally, we can estimate the derivatives of γ in t by differentiating (1.2).

Proposition 1.4 (uniqueness) *Let γ_1 and γ_2 be two solutions of (1.2) in $\widetilde{C}^{0,1}(S^1 \times [0,T))$ where F is parabolic. If $\gamma_1(\cdot,0) = \gamma_2(\cdot,0)$ in some parametrization, then $\gamma_1(\cdot,t) = \gamma_2(\cdot,t)$ for all t.*

Proof: Represent γ_1 and γ_2 as graphs over a smooth curve close to $\gamma_1(\cdot,0)$ and then use Fact 6 to deduce that they are identical for all t in $[0,T)$. Note that representing a closed curve as a graph over some smooth curve is possible provided the curve is Lipschitz continuous. □

Now we formulate two useful properties of the flow (1.2). They are immediate consequences of the strong maximum principle.

Proposition 1.5 (preserving embeddedness)   *Consider (1.2) where F is parabolic and symmetric. Then, any solution $\gamma(\cdot,t)$ in $\widetilde{C}^{0,1}(S^1 \times [0,T))$ is embedded if γ_0 is embedded.*

Proof: Since $\gamma(\cdot,t)$ tends to γ_0 in $C^{0,1}$-norm as $t \downarrow 0$, $\gamma(\cdot,t)$ is embedded for all small t. Suppose that $\gamma(\cdot,t)$ has a self-intersection at some time. We can find $p, q, p \neq q$, and $t_1 > 0$ such that $\gamma(p,t_1) = \gamma(q,t_1)$, but $\gamma(\cdot,t)$ is embedded for all t less than t_1. Without loss of generality we may take $\gamma(p,t_1) = (0,0)$ and $\boldsymbol{n}(p,t_1) = (0,1)$. In a sufficiently small rectangle $R = (-a,a) \times (-b,b)$, $R \bigcap \gamma(\cdot,t_1)$ is the union of two graphs $u_i, i = 1,2$, over $(-a,a)$ satisfying $u_1(0,t_1) = u_2(0,t_1)$, $u_{1x}(0,t_1) = u_{2x}(0,t_1)$ and $u_2 \geqslant u_1$ at $t = t_1$. Moreover, $u_2 > u_1$ for all $t < t_1$ and close to t_1. By the embeddedness of $\gamma(\cdot,t), t < t_1$, $(x, u_2(x,t))$, say, is transversed along the negative x-axis and $(x, u_1(x,t))$ along the positive x-axis. However, since F is symmetric, reversing the direction of $(x, u_2(x, t))$ does not change the equation it satisfies. So both u_1 and u_2 solve the same equation, and we may use the strong maximum principle to conclude that $u_2 = u_1$ for all $t < t_1$, and the contradiction holds.   □

Proposition 1.6 (strong separation principle) *Consider (1.2) where F is parabolic and symmetric. Let γ_1 and γ_2 be two solutions of (1.2) in $C([0,T]; C^{0,1}(S^1))$ and let $D_1(t)$ be a component of $\mathbb{R}^2 \setminus \gamma_1(\cdot, t)$ whose boundary $\partial D_1(t)$ changes continuously in t. Suppose that $\gamma_2(\cdot, 0)$ is contained in $\overline{D_1(0)}$ and touches $\partial D_1(0)$ only at regular points of $\gamma_1(\cdot, 0)$ or is disjoint from $\partial D_1(0)$. Then, $\gamma_2(\cdot, t)$ is contained in $D_1(t)$ for all $t > 0$.*

A point on a curve γ is called a regular point if there is a disk D containing this point such that $D \bigcap \gamma(\cdot)$ is a one-dimensional manifold.

Proof: Parametrize both $\gamma_1(\cdot, 0)$ and $\gamma_2(\cdot, 0)$ by arc-length. We shall prove the proposition when $\gamma_2(\cdot, 0) \bigcap \overline{\partial D_1(0)}$ is non-empty. Let X be a point on $\gamma_2(\cdot, 0) \bigcap \overline{\partial D_1(0)}$. By our hypotheses, we can find a maximal interval $[-a, a]$ on which $\gamma_2(\cdot, 0)$ and $\gamma_1(\cdot, 0)$ coincide. Now represent $\gamma_2(\cdot, 0)$ and $\gamma_1(\cdot, 0)$ as graphs over a smooth curve defined in $J = [-a - \delta, a + \delta]$ for some small δ. We may assume that $d_2(\cdot, 0) \geqslant d_1(\cdot, 0)$ on J and $d_2(\pm(a + \delta), t) > d_2(\pm(a + \delta), t)$ for all small t. By Fact 6, $d_2 > d_1$ for $t > 0$. In other words, γ_2 is separated from γ_1 instantly. A similar argument shows that they cannot touch afterward.   □

At this point, we present an example showing that the flow (1.2) in general, depends on the parametrization of the initial curve. Let C be the circle $\{x^2 + y^2 = 4\}$ and C' the circle $\{x^2 + (y + 1)^2 = 1\}$. We may modify both circles near $A = (0, -2)$ so that the resulting curves have smooth contact at A. Let $B = (0, 2)$ and $B' = (0, 0)$. Consider two curves: γ_1 is $ABAB'AB'A$ and γ_2 is $ABABAB'A$, both parametrized by arc-length and transversing in counterclockwise direction. Mark a point P and a point Q on C which lie on the left and the right of A, respectively. Consider the arc $PABAQ$ on γ_2. We can find a corresponding arc $P'ABAQ'$ on γ_1 with the same length. When P and Q are close to A, we can represent the second arc as a graph over the first arc. So $d_1(\cdot, 0) \geqslant d_2(\cdot, 0) \equiv 0$ and $d_1(\cdot, t) > d_2(\cdot, t)$ for small t at the endpoints. By Fact 6, the arcs will be separated instantly. It shows that, although the geometry of γ_1 and γ_2 are the same, their flows are different.

Now we give some consequences of the Sturm oscillation theorem (see Fact 7 in the next section).

Proposition 1.7 *Let γ_1 and γ_2 be two solutions of (1.2) in $C^{\infty}(S^1 \times (0, T))$ where F is parabolic and symmetric. Denote the number of intersection points of $\gamma_1(\cdot, t)$ and $\gamma_2(\cdot, t)$ by $Z(t)$. Then $Z(t)$ is finite for all t in $(0, T)$, and it drops exactly at those instants t when $\gamma_1(\cdot, t)$ and $\gamma_2(\cdot, t)$ touch tangentially at some point. Moreover, all these instants form a discrete subset of $(0, T)$.*

Proof: Use the local form (1.3) and Fact 7 in the next section. □

Proposition 1.8 *Let γ be a solution of (1.2) in $C^{\infty}(S^1 \times (0, T))$ where F is parabolic. Denote the number of nodes of $\gamma(\cdot, t)$ by $N(t)$. Then $N(t)$ is finite for all t in $(0, T)$ and drops exactly at those instants when $F(\gamma(\cdot, t), \theta(\cdot, t), k(\cdot, t))$ has a double zero. The instants at which this happens form a discrete set.*

A point $\gamma(p)$ on a curve γ is a **node** of F if $F(\gamma(p), \theta(p), k(p)) = 0$. In the curve shortening flow, a node is simply an inflection point of the curve.

Proof: For each fixed t^*, we may represent $\gamma(\cdot, t)$ as graphs over $\gamma(\cdot, t^*)$ for all t close to t^*. From (1.4), we know that a node of $\gamma(\cdot, t)$ is a zero of $d_t(\cdot, t)$. By differentiating (1.4), d_t is seen to satisfy a linear parabolic equation, and so the proposition follows from Fact 7.   □

We end this section by establishing a more general existence result for immersed curves.

Proposition 1.9 *Consider the Cauchy problem for (1.2) where F is uniformly parabolic, symmetric, and satisfies (i) for any compact K in $\mathbb{R}^2 \times S^1 \times \mathbb{R}$, there exists a constant C such that*

$$\left|\frac{\partial F}{\partial x}\right| + \left|\frac{\partial F}{\partial y}\right| + \left|\frac{\partial F}{\partial \theta}\right| + \left|\frac{\partial F}{\partial q}\right| |q| \leqslant C(1 + q^2) , \qquad (1.8)$$

for all (x, y, θ, q) in $K \times \mathbb{R}$, and (ii)

$$F(\gamma, \theta, 0) \ only \ depends \ on \ \theta . \qquad (1.9)$$

Then, for any closed $C^{0,1}$-curve γ_0, there exists $T > 0$ such that the Cauchy problem for (1.2) has a unique solution in $\widetilde{C}^{0,1}(S^1 \times [0, T)) \cap C^\infty(S^1 \times (0, T))$.

Proof: Since γ_0 is a $C^{0,1}$-curve, we can find open intervals $J_\alpha, \alpha = 1, \cdots, N$, in different coordinates and $C^{0,1}$-functions u_0^α defined over J_α such that γ_0 can be described as the union of the graphs $\{(x, u_0^\alpha(x)) : x \in J_\alpha\}$, $\alpha = 1, \cdots, N$. Let $\{\gamma_0^j\}$ be a smooth approximation to γ_0 where γ_0^j is the graph of some smooth $u_0^{\alpha,j}$ in J_α converging to u_0^α in $C^{0,1}$-norm. By Proposition 1.2, for each j there is a maximal solution γ^j taking γ_0^j as its initial curve in $[0, \omega^j), \omega_j > 0$. We first show that $\{\omega^j\}$ has a uniform positive lower bound.

Let C_0 be any circle of radius δ centered at a point lying on the 3δ-neighborhood of γ_0 and let $C(t)$ be the flow (1.2) starting at

C_0. There is a uniform, positive t_1 such that all $C(t)$ exist and stay outside the δ-neighborhood of γ_0 for all t in $[0, t_1]$. By the strong separation principle, we know that, for sufficiently large j, $\gamma^j(\cdot, t)$ is confined to the δ-neighborhood of γ_0 for $t < \min\{\omega^j, t_1\}$. On the other hand, for each α and $x \in J_\alpha$, consider the vertical line segment ℓ_x which passes the point $(x, 0)$ and whose endpoints lie on the boundary of the 4δ-neighborhood of γ_0. Since $F(\gamma, \theta, 0)$ only depends on θ, the flow (1.2) starting at ℓ_x, $\ell_x(t)$, is again a vertical line segment, and it moves horizontally at constant speed. For each α, fix a subinterval J'_α, $\overline{J'_\alpha} \subseteq J_\alpha$, so that the union of the graphs $(x, u^{\alpha,j}(x, t))$ over J'_α still covers to $\gamma^j(\cdot, t)$. There exists $t_2 > 0$ such that the line segments $\ell_x(t), x \in J_\alpha$, cover J'_α for all $t < t_2$. Besides, the endpoints of $\ell_x(t)$ lie outside the 2δ-neighborhood of γ_0. Now, as each ℓ_x intersects the graph of $u_0^{\alpha,j}$ transversally at exactly one point, by the Sturm oscillation theorem (Fact 7 in the next section) the graph of $u^{\alpha,j}(\cdot, t)$ cannot become vertical in $J'_\alpha \times [0, t_3]$, $t_3 = \min\{\omega^j, t_1, t_2\}$. This means that the gradient of $u^{\alpha,j}$ is bounded. Observe that $u \equiv u^{\alpha,j}$ satisfies (1.3), which is now written as

$$u_t = \sqrt{1 + u_x^2}\ \Phi(x, u, u_x, u_{xx}) \ .$$

So $w \equiv \partial u^{\alpha,j}/\partial x$ satisfies

$$w_t = (\sqrt{1 + w^2}\ \Phi(x, u(x, t), w, w_x))_x \ .$$

By uniform parabolicity and the structural condition (1.7), it follows from Theorem 3.1 in Chapter 5 of [86] that w_x is uniformly bounded on every compact subset of $J'_\alpha \times (0, t_3]$. Thus, the curvature of each γ^j is bounded as long as $t \leqslant t^* = \min\{t_1, t_2\}$. By Proposition 1.2, γ^j exist in $[0, t^*]$ for all large j.

Next, we derive a uniform gradient estimate for $u^{\alpha,j}$ by the following trick. Instead of using ℓ_x, we tilt it a little bit to the right and to the left to obtain line segments ℓ'_x and ℓ''_x. Each ℓ'_x (resp. ℓ''_x) makes an angle $\theta' < \pi/2$ (resp. $\theta'' > \pi/2$) with the positive x-axis. When θ' and θ'' are close to $\pi/2$, ℓ'_x and ℓ''_x are still transversal to the graph of $u_0^{\alpha,j}$, and their endpoints still lie outside the δ-neighborhood of γ_0. By the Sturm oscillation theorem again,

$$|u_x| \leqslant \max\{\tan\theta', |\tan\theta''|\}$$

for all $u = u^{\alpha,j}$ over $J'_\alpha \times [0, t^*]$. As before, it implies a uniform gradient bound on every compact subset of $J'_\alpha \times (0, t^*]$ and, by parabolic regularity, all higher order bounds follow. By passing to a convergent subsequence, we obtain a solution of (1.2) which is a local Lipschitz graph of $u^\alpha(x,t)$, $x \in J'_\alpha, t \leqslant t^*$ with uniform $C^{0,1}$-norm. Proposition 1.10 below ensures that this solution takes γ_0 as its initial curve and it belongs to $\tilde{C}^{0,1}(S^1 \times [0, t^*])$. □

Proposition 1.10 *Let γ be a solution of (1.2) in $[0,T)$ where F is uniformly parabolic and symmetric. Suppose that $\{\gamma(\cdot,t) : t \in [0,T)\}$ is contained in some bounded set. There exist positive constants ρ and C depending on F and the bounded set such that*

$$\gamma(\cdot,t) \subseteq N_{C\sqrt{t-t_0}}(\gamma(\cdot,t_0)) \tag{1.10}$$

for all t_0 and t satisfying $0 \leqslant t - t_0 < \min\{\rho, T\}$.

Proof: By uniform parabolicity, there exist positive constants k_0 and C_1 such that

$$\sup\{F(x,y,\theta,q) : (x,y) \in D \text{ and } \theta \in \mathbb{R}\} \leqslant C_1 q$$

for all $q \geqslant k_0$, where D is a bounded set containing the image of $\gamma(\cdot,t)$, for all $t \in [0,T)$. Let C_0 be any circle of radius δ, $\delta \leqslant 1/k_0$, centered at the boundary of the 2δ-neigborhood of $\gamma(\cdot,t), t_0 \in [0,T)$, and denote the solution of

$$\frac{\partial C}{\partial t} = C_1 k\boldsymbol{n}, \qquad C(\cdot,t_0) = C_0 ,$$

by $C(t)$. This solution is a family of shrinking circles whose radius at t is given by $\sqrt{\delta^2 - 2C_1 t}$. Hence, it exists for t in $[t_0, t_0 + \delta^2/(2C_1))$. Noticing that the curvature of $C(t)$ is increasing and so (1.10) continues to hold, we infer from Proposition 1.6 that $\gamma(\cdot, t)$ and $C(t)$ are disjoint. Since the center of $C(t)$ could be any point on the boundary of $N_{2\delta}(\gamma(\cdot, t_0))$, $\gamma(\cdot, t)$ stays inside $N_{2\delta}(\gamma(\cdot, t_0))$ as long as $0 \leqslant t - t_0 < \delta^2/(2C_1)$. In particular, taking $t - t_0 = \delta^2/(4C_1)$ yields the desired result where $C = (\delta C_1)^{1/2}$ and $\rho = (4C_1 k_0^2)^{-1}$. □

1.2 Facts from the parabolic theory

Let E be a region in $\mathbb{R}^2$. We consider the equation

$$u_t = \Phi(x, t, u, u_x, u_{xx}) , \quad (x, t) \in E , \tag{1.11}$$

where $\Phi(x, t, z, p, q)$ is smooth in $\overline{E} \times \mathbb{R}^3$. This equation is parabolic if $\partial \Phi/\partial q$ is positive in $E \times \mathbb{R}^3$, and uniformly parabolic if $\lambda \leqslant \partial \Phi/\partial q \leqslant \Lambda$ holds for some positive constants λ and Λ. A function defined in E is called a (classical) solution of (1.11) if it is continuously differentiable in t, twice continuously differentiable in x and satisfies (1.11). From now on, a solution always means a classical solution.

The parabolic Hölder space is defined as follows. Let $Q = I \times (0, T)$ be a cylinder. For a function u defined in Q, we introduce the following semi-norms and norms: for each integer $k \geqslant 0$ and $\alpha \in (0, 1]$, set

$$[u]_\alpha = \sup \left\{ \frac{|u(x, t) - u(y, s)|}{(|x - y|^2 + |t - s|)^{\alpha/2}} : (x, t), (y, s) \in Q \right\} ,$$

$$\|u\|_{\widetilde{C}^{k,\alpha}(\overline{Q})} = \sum_{i+2j \leqslant k} \left\| \frac{\partial^{i+j} u}{\partial x^i \partial t^j} \right\|_{C(\overline{Q})} + \sum_{i+2j=k} \left[\frac{\partial^{i+j} u}{\partial x^i \partial t^j} \right]_\alpha .$$

The parabolic Hölder space $\widetilde{C}^{k,\alpha}(\overline{Q})$ is the completion of $C^{\infty}(\overline{Q})$ under $\|\cdot\|_{\widetilde{C}^{k,\alpha}(\overline{Q})}$.

We first state results for the linear equation

$$u_t = a(x,t)u_{xx} + b(x,t)u_x + c(x,t)u - f(x,t) \ . \qquad (1.12)$$

Fact 1 (*a priori* **estimates**) *Consider (1.12) in $S^1 \times (0,T)$ where it is uniformly parabolic and the coefficients and f are 2π-periodic and in $\widetilde{C}^{k,\alpha}(\overline{Q})$ for some (k,α). Then, for any solution u in $\widetilde{C}^{k+2,\alpha}(\overline{Q})$, there exists a constant C which depends on $\lambda, \Lambda, k, \alpha$ and the $\widetilde{C}^{k,\alpha}$-norms of the coefficients such that*

$$\|u\|_{\widetilde{C}^{k+2,\alpha}(\overline{Q})} \leqslant C\Big(\|f\|_{\widetilde{C}^{k,\alpha}(\overline{Q})} + \|u(\cdot,0)\|_{C^{k+2,\alpha}(S^1)}\Big) \ .$$

Moreover, for any sub-cylinder $Q' = S^1 \times (t_0,T)$, $t_0 > 0$ there exists a constant C' which depends on $t_0, \lambda, \Lambda, k, \alpha$ and the $\widetilde{C}^{k,\alpha}$-norms of the coefficients such that

$$\|u\|_{\widetilde{C}^{k+2,\alpha}(\overline{Q'})} \leqslant C'\Big(\|f\|_{\widetilde{C}^{k,\alpha}(\overline{Q})} + \|u(\cdot,0)\|_{C(S^1)}\Big) \ . \qquad (1.13)$$

Fact 2 (existence and uniqueness) *Consider (1.12) in $S^1 \times (0,T)$ where it is uniformly parabolic, and the coefficients and f are 2π-periodic and in $\widetilde{C}^{k,\alpha}(\overline{Q})$ for some (k,α). Then, for any u_0 in $C^{k+2,\alpha}(S^1)$, there exists a unique solution u in $\widetilde{C}^{k+2,\alpha}(\overline{Q})$ satisfying $u(\cdot,0) = u_0$.*

Fact 2 may be proved in the following way. First, solve the Cauchy problem in the smooth category by means of separation of variables. Then use an approximation argument coupling with the global *a priori* estimate in Fact 1 to get the general result.

Now we consider the nonlinear equation (1.11).

Fact 3 (local solvability) *Consider (1.11) in $S^1 \times (0, T)$ where Φ is smooth and parabolic. For any u_0 in $C^{k+2,\alpha}(S^1)$ for some (k, α), there exists a positive $t_0 \leqslant T$ such that (1.11) has a solution u in $\tilde{C}^{k+2,\alpha}(\overline{Q})$ satisfying $u(\cdot, 0) = u_0$. Moreover, if u_0 depends smoothly on a parameter (resp. analytically on a parameter and Φ is analytic), then u also depends smoothly (resp. analytically) on the same parameter.*

Fact 3 is a consequence of the inverse function theorem. In fact, by replacing u by $u - u_0$, we may assume the initial value is identically zero. Let $Q(t) = S^1 \times (0, t)$, $t \leqslant t_0$ where t_0 is to be chosen later, and define a map $\mathcal{F}$ from $X = \{u \in \tilde{C}^{k+2,\alpha}(\overline{Q(t)}) : u(\cdot, 0) = 0\}$ to $\tilde{C}^{k,\alpha}(\overline{Q(t)})$ by

$$\mathcal{F}(u) = u_t - \Phi(x, t, u, u_x, u_{xx}) .$$

The Fréchet derivative of $\mathcal{F}$ at the specified function $u^0 = \Phi(x, t, 0, 0, 0)t$ is given by

$$DF(u^0)v = v_t - \left(\frac{\partial \Phi}{\partial q} v_{xx} + \frac{\partial \Phi}{\partial p} v_x + \frac{\partial \Phi}{\partial z} v\right) ,$$

where the coefficients are evaluated at u^0. Since Φ is parabolic, it follows from Fact 2 that $DF(u^0)$ is invertible. By the inverse function theorem there exist t_0, ρ, δ such that, for any f satisfying $\|f - \mathcal{F}(u^0)\|_{\tilde{C}^{k,\alpha}(\overline{Q(t)})} < \delta$, there exists a unique u satisfying $\|u - u^0\|_{\tilde{C}^{k+2,\alpha}(\overline{Q(t)})} < \delta$ such that $\mathcal{F}(u) = f$ for all $t \leqslant t_0$. As $\mathcal{F}(u^0)$ tends to zero as $t \downarrow 0$, there is some $t_1 > 0$ such that $\|\mathcal{F}(u^0)\|_{\tilde{C}^{k,\alpha}(\overline{Q(t)})} < \delta$. In other words, $\mathcal{F}(u) = 0$ is solvable in $\tilde{C}^{k+2,\alpha}(S^1 \times [0, t_1])$. Now, the smooth\analytic dependence on a parameter is a consequence of the implicit function theorem. For an appropriate version see, e.g., Zeidler [112].

Fact 4 (instant smoothness\analyticity) *Let $u \in \tilde{C}^{2,\alpha}(\overline{Q})$ be a solution of (1.11) where Φ is smooth and parabolic. Then u belongs to $C^{\infty}(S^1 \times (0,T))$. Moreover, if Φ is further analytic, then u is analytic in $S^1 \times (0,T)$.*

This fact can be deduced from Fact 3. For any solution u, we set $u_{a,b} = u(x + at, bt)$ where a is near 0 and b near 1. Then $u_{a,b}$ solves (1.11) for $\Phi = \Phi_{a,b} \equiv au_x + b\Phi(x + at, bt, u, u_x, u_{xx})$, which depends smoothly\analytically on (a,b). According to Fact 3,

$$\left. \frac{\partial^{j+k} u_{a,b}}{\partial a^j \partial b^k} \right|_{(a,b)=(0,1)} = t^{j+k} \left. \frac{\partial^{j+k} u}{\partial x^j \partial t^k} \right|_{(x,t)}.$$

Hence, u is smooth\analytic for $t > 0$.

Next we consider the strong maximum principle. Recall that a (classical) subsolution (resp. supersolution) of (1.11) or (1.12) is a function which satisfies (1.11) or (1.12) with "=" replaced by "$\leqslant$" (resp. "$\geqslant$"). For any (x_0, t_0) in E, we define $E(x_0, t_0)$ to be the set consisting of all points in E which lie on or below the horizontal line segment ℓ passing through (x_0, t_0) in E or can be connected to ℓ by a vertical line segment contained inside E.

Fact 5 (strong maximum principle) *Consider (1.12) where it is uniformly parabolic, the coefficients are bounded, $c \leqslant 0$ and $f = 0$. Suppose u is a subsolution of (1.12) which attains a non-negative maximum M at (x_0, t_0) in E. Then $u = M$ in $E(x_0, t_0)$.*

Fact 6 (strong comparison principle) *Let u and v be, respectively, a subsolution and a supersolution of (1.11) which is parabolic, in $C(\overline{Q})$. Suppose that $u \leqslant v$ on the parabolic boundary $\partial_p Q \equiv \overline{I} \times \{0\} \bigcup \partial I \times [0,T)$. Then, either $u \equiv v$ or $u < v$ in $\overline{Q} \setminus \partial_p Q$.*

Let $w = e^{\lambda t}(u - v)$. Then, for sufficiently large λ, w satisfies a uniformly parabolic, linear equation of the form (1.12) where $c \leqslant 0$. So Fact 6 follows from Fact 5. In Chapter 8 we need a weak version of Fact 6. Let's consider (1.12) where a is non-negative, the coefficients are bounded and $f \equiv 0$. The **weak maximum principle** states that a subsolution which is non-positive on the parabolic boundary of E is non-positive in E. Using this principle, we immediately deduce the **weak comparison principle**: Let u and v be a subsolution and a supersolution of (1.11) in $C(\overline{Q})$, respectively. Suppose that $\partial\Phi/\partial q \geqslant 0$ and $u \leqslant v$ on $\partial_P Q$. Then $u \leqslant v$ in $\overline{Q}$.

Finally, we need the following:

Fact 7 ("Sturm oscillation theorem") *Consider (1.12) in* Q *where it is uniformly parabolic,* $a, a_x, a_{xx}, a_t, b, b_x, b_t$ *and* c *are bounded measurable, and* $f = 0$. *Suppose* u *is a solution which never vanishes on* $\partial I \times [0, T]$. *Then* $Z(t)$, *the number of zeroes of* $u(\cdot, t)$, *is finite and non-increasing for all* $t \in (0, T)$ *and it drops exactly at multiple zeroes. The set* $\{t \in (0, T) : u(\cdot, t) \text{ has a multiple zero }\}$ *is discrete.*

1.3 The evolution of geometric quantities

In this section, we derive the evolution equations satisfied by the basic geometric quantities for the solution of (1.6). We first compute the evolution of the arc-length element $s = |\gamma_p|$. By (1.1), we have

$$
\begin{aligned}
\frac{\partial s^2}{\partial t} &= 2\left\langle \frac{\partial \gamma}{\partial p}, \frac{\partial}{\partial t}\frac{\partial \gamma}{\partial p} \right\rangle \\
&= 2\left\langle \frac{\partial \gamma}{\partial p}, \frac{\partial}{\partial p}\frac{\partial \gamma}{\partial t} \right\rangle
\end{aligned}
$$

$$= -2Fs^2k + 2s^2G_s.$$

Hence

$$\frac{\partial s}{\partial t} = (-Fk + G_s)s .\tag{1.14}$$

Next, the unit tangent t satisfies

$$\frac{\partial t}{\partial t} = -\frac{1}{s^2}\frac{\partial s}{\partial t}\frac{\partial \gamma}{\partial p} + \frac{1}{s}\frac{\partial}{\partial p}(Fn + Gt)$$

$$= (F_s + Gk)n.$$

Now,

$$\frac{\partial n}{\partial t} = \left\langle \frac{\partial n}{\partial t}, n \right\rangle n + \left\langle \frac{\partial n}{\partial t}, t \right\rangle t$$

$$= -(F_s + Gk)t.$$

By differentiating both sides of $t = (\cos\theta, \sin\theta)$ we have

$$\frac{\partial \theta}{\partial t} = F_s + Gk.\tag{1.15}$$

By the Frenet formulas, we know that $k = \partial\theta/\partial s$. Therefore,

$$\frac{\partial k}{\partial t} = \frac{\partial}{\partial t}\left(\frac{1}{s}\frac{\partial \theta}{\partial p}\right)$$

$$= -\frac{1}{s^2}\frac{\partial s}{\partial t}\frac{\partial \theta}{\partial p} + \frac{1}{s}\frac{\partial}{\partial p}\left(\frac{\partial \theta}{\partial t}\right)$$

$$= (Fk - G_s)k + (F_s + Gk)_s.\tag{1.16}$$

In particular,

$$\frac{\partial k}{\partial t} = k_{ss} + k^3\tag{1.17}$$

holds for the CSF.

The length of $\gamma(\cdot, t), L(t)$, satisfies

$$
\begin{aligned}
\frac{dL}{dt} &= \int_I \frac{\partial s}{\partial t}\, dp \\
&= -\int (Fk - G_s)ds \qquad\qquad (1.18) \\
&= -\int Fk\, ds.
\end{aligned}
$$

So, the curve shortening flow is the negative L^2-gradient flow for the length. For the generalized curve shortening flow for closed curves, we have

$$
\frac{dL}{dt} = -\int_\gamma |k|^{1+\sigma} ds.
$$

So, the length of the curve is strictly decreasing along the flow. In fact, by

$$
\int_\gamma |k| ds \geqslant 2\pi
$$

and the Hölder inequality, we have

$$
\frac{dL}{dt} \leqslant -(2\pi)^{\sigma+1} L^{-\sigma} ,
$$

and

$$
L^{1+\sigma}(t_0) \leqslant L^{1+\sigma}(0) - (2\pi)^{1+\sigma}(1+\sigma)t .
$$

It implies an upper bound on the life span which only depends on the length of the initial curve.

For an embedded closed solution we can use the formula for the enclosed area

$$A = -\frac{1}{2} \int_\gamma < \gamma, \boldsymbol{n} > ds$$

to compute

$$\frac{dA}{dt} = -\frac{1}{2} \int F \, ds + \frac{1}{2} \int \left(F_s + Gk \right) < \gamma, \boldsymbol{t} > ds$$

$$+ \frac{1}{2} \int \left(Fk - G_s \right) < \gamma, \boldsymbol{n} > ds \ ,$$

which, after integration by parts, gives

$$\frac{dA}{dt} = - \int_\gamma F \, ds \ . \tag{1.19}$$

For the curve shortening flow, we have the following remarkable property:

$$A(t) = A(0) - 2\pi t \ . \tag{1.20}$$

It says that, no matter what the initial curve is, the life span of the flow is equal to $A(0)/2\pi$, which only depends on the initial area. (In the next chapter, we shall show that the flow exists until $A(t) = 0$.)

Notes

Local existence. It was Gage and Hamilton [58] who first pointed out the weak parabolicity of the curve shortening flow when it is viewed as a system for (γ^1, γ^2). Instead of reducing it to a single parabolic equation, they employed a version of the Nash-Moser implicit function theorem to show local existence. However, in subsequent work, local existence was obtained via reduction to a single

equation in various way. In [13] and [14], Angenent studied (1.2) on
a surface. Very general results on local existence and the existence
of limit curves are obtained. In particular, Propositions 1.2–1.10 are
essentially contained in [14]. Some of his main results on local existence may be described as follows.

Let M be a two-dimensional Riemannian manifold and let $\gamma(\cdot, t)$
be a family of curves, from S^1 to M. At each point of the curve,
we can decompose the velocity vector γ_t into its normal and tangent
components with respect to a frame $(\boldsymbol{t}, \boldsymbol{n})$. Consider the flow

$$V = F(\boldsymbol{t}, k) , \tag{1.21}$$

where V is the normal velocity of $\gamma(\cdot, t)$. The function F is defined
in $S^1(M) \times \mathbb{R}$ where $S^1(M)$ is the unit tangent bundle of M and $\boldsymbol{t}, k$
are, respectively, the tangent and curvature of $\gamma(\cdot, t)$. For simplicity,
F is assumed to be smooth in $S^1(M) \times \mathbb{R}$, and it further satisfies
some of the following assumptions: for some positive λ, Λ, μ and ν,

(i) $\lambda \leqslant \partial F / \partial k \leqslant \Lambda$,

(ii) $\left| F(\boldsymbol{t}, 0) \right| \leqslant \mu$,

(iii) $\left| \nabla^h F \right| + \left| k \nabla^v V \right| \leqslant \nu \left(1 + |k|^2 \right)$,   and

(iv) $F(-\boldsymbol{t}, -k) = -F(\boldsymbol{t}, k)$,

in $S^1(M) \times \mathbb{R}$. Here ∇^h and ∇^v are, respectively, the horizontal and
vertical components of the gradient of F, ∇F, in $\boldsymbol{t}$. Now we can
state:

Theorem A *Under (i)–(iii), the Cauchy problem for (1.21) has a
maximal solution in $(0, \omega), \omega > 0$, for any locally Lipschitz γ_0.*

Theorem B *Under (i)–(vi), the Cauchy problem for (1.21) has a
maximal solution in $(0, \omega), \omega > 0$, for any C^1-locally graph-like curve
γ_0.*

See [14] for the definition of a C^1-locally graph-like curve. We point out that any curve which is locally the graph of a continuous function is C^1-locally graph-like. In particular, it implies that Proposition 1.9 holds without (1.8).

Parabolic theory. We refer to the books [86], Protter-Weinberger [96], and Lieberman [88] for detailed information on the basic facts of parabolic theory, except for Fact 7. The deduction of Fact 4 from Fact 3 is taken from [13] where it is attributed to DaPrato and Grisvard. Fact 7, the most general form of the Sturm oscillation theorem for parabolic equations, is taken from Angenent [12]. Results of this type were established by many authors since Sturm in 1836.

Geometric flows. The equivalence between (1.2) and (1.6) was formulated in Epstein-Gage [48]. Here our proof follows Chou [29]. Usually, people choose G to vanish identically. Another useful choice is to make the parametrization have constant speed at each t. We shall illustrate this point below.

Apparently, Proposition 1.1 holds when F and G depend also on the derivatives of k with respect to the arc-length. Equation (1.2) in this form may be called a *geometric evolution equation*. Quite a number of geometric evolution equations have been studied in recent years. Usually, they are closely related to geometric functionals, such as the area, the length, or some curvature integrals. For instance, the generalized curve shortening flow decreases both the area and the length of an embedded, closed curve. In some physical situations, one would like to have a flow which decreases the length but preserves the area (the total mass). The simplest choice of parabolic flows of this kind is the flow driven by surface diffusion

$$\frac{\partial \gamma}{\partial t} = -k_{ss} n \ .$$

The reader is referred to Cahn-Taylor [24], Cahn-Elliott-Novick-Cohen [23] and Giga-Ito [64] for physical background and analysis of this flow.

Flows decreasing the elastic energy

$$\int k^2 ds$$

are also studied by some authors. By taking variations in different function spaces, one arrives at three different curvature flows associated to this energy, see Langer-Singer [87], Wen [110], and [111].

When one does not insist on parabolic equations and looks for flows which preserve length and area, one may consider F depending on the derivatives of the curvature in odd orders. As the simplest example, let's take $F = -k_s$. Clearly, the resulting flow preserves length and area. Let $\gamma(\cdot, t)$ be a solution of this flow. We reparametrize it by setting $\gamma'(\cdot, t) = \gamma(p(s, t), t)$ where $s = s(p, t)$ is the arc-length element of $\gamma(\cdot, t)$. Then $\gamma'(\cdot, t)$ satisfies

$$\frac{\partial \gamma'}{\partial t} = -k_s n + G t \ .$$

G can be determined by requiring $\partial s / \partial t = 0$. On one hand, we have

$$\gamma'_{ts} = \left(-k_{ss} + Gk \right) n + \left(k_s k + G_s \right) t$$

by (1.1). On the other hand, we have

$$\gamma'_{st} = \left(-k_{ss} + Gk \right) n \ ,$$

by (1.15). Since $\partial s / \partial t = 0$, we have $\gamma'_{ts} = \gamma'_{st}$. It implies that $G = -k^2/2$, and so the flow is given by

$$\gamma'_t = -k_s n - \frac{1}{2} k^2 t \ . \tag{1.22}$$

Now, by (1.16), the curvature $k(s,t)$ satisfies the modified KdV equation,

$$k_t + k_{sss} + \frac{3}{2}k^2 k_s = 0 \ . \tag{1.23}$$

This equation is one among many integrable equations studied extensively in the past several decades. Since a curve is determined by its curvature up to a rigid motion, there is a formal equivalence between (1.22) and (1.23). It turns out that many integrable equations are associated to the motion of curves in the plane or the space in this way. For further discussion, see Goldstein-Petrich [69] and Nakayama-Segur-Wadati [91].

Chapter 2

Invariant Solutions for the Curve Shortening Flow

We discuss invariant solutions – travelling waves, spirals and self-similar solutions – for the generalized curve shortening flow (GCSF)

$$\frac{\partial \gamma}{\partial t} = |k|^{\sigma-1} k \boldsymbol{n} \ , \quad \sigma > 0 \ . \tag{2.1}$$

They are not merely examples. In fact, they can be used as comparison functions to yield *a priori* estimates. Even more significant is their role in the classification of the singularities of (2.1). For the curve shortening flow, the travelling waves, which are called grim reapers, describe the asymptotic shape of type II singularity and the contracting self-similar solutions (they are circles when embedded) characterize type I singularities. We shall discuss this in Chapter 5. Travelling waves, spirals, and expanding self-similar solutions appear to be very stable. They can be used to describe the long time behaviour of complete, unbounded solutions of (2.1).

2.1 Travelling waves

A travelling wave is a solution of (2.1) which assumes the form $v(x,t) = v(x) + ct$ when it is expressed as a graph locally over some

x-axis. It satisfies the equation

$$v_{xx} = \left(1 + v_x^2\right)^{\frac{3\sigma-1}{2\sigma}} , \qquad (2.2)$$

after a scaling in (x, t) to take $c = 1$. It is easy to see that any solution of (2.2) is even and convex. If we rotate the axes by $90°$, the graph $(x, v(x))$ consists of two branches of the form $x^{\pm}(v - t)$ and thus justifies the name of a travelling wave.

Depending on the value of σ, these solutions can be put into two classes. First, for $0 < \sigma \leqslant 1/2$, v is entire over $\mathbb{R}$ and satisfies

$$\lim_{x \longrightarrow \infty} \frac{v(x)}{x^{\frac{1-\sigma}{1-2\sigma}}} = \left(\frac{1 - 2\sigma}{\sigma}\right)^{\frac{1-\sigma}{1-2\sigma}} \left(\frac{\sigma}{1 - \sigma}\right) , \qquad \sigma < \frac{1}{2} ,$$

and

$$\lim_{x \longrightarrow \infty} \frac{v(x)}{e^x} = \frac{1}{2} , \qquad \left(\sigma = \frac{1}{2}\right) .$$

When $\sigma > 1/2$, there exists a finite $\bar{x}$ depending on σ such that dv/dx blows up at $\bar{x}$. Moreover,

$$\lim_{x \uparrow \bar{x}} \frac{v(x)}{-\log(\bar{x} - x)} = 1 \quad (\sigma = 1) ,$$

$$\lim_{x \uparrow \bar{x}} \frac{v(x)}{(\bar{x} - x)^{\frac{1-\sigma}{1-2\sigma}}} = \left(\frac{2\sigma - 1}{\sigma}\right)^{\frac{1-\sigma}{1-2\sigma}} \left(\frac{\sigma}{1 - \sigma}\right) \quad \left(\frac{1}{2} < \sigma < 1\right) ,$$

and

$$\lim_{x \uparrow \bar{x}} v(x) = \bar{v} < \infty , \quad (\sigma > 1) .$$

It is worthwhile to point out that when $\sigma = 1$, $v(x)$ is given explicitly by the "grim reaper"

$$v(x) = \log \sec x + \text{ constant} , \quad x \in (-\pi/2 , \pi/2) .$$

When $\sigma > 1$, travelling waves are not complete as curves.

Travelling waves with speed c are given by the graphs of $c^{-1}v(cx)$. Intuitively speaking, the speed increases as the width of the wave narrows upward.

2.2   Spirals

Spirals are travelling waves in the polar angle α. To describe its equation, we express the solution curve as $\big(r\cos(\alpha(r)+ct),\, r\sin(\alpha(r)+ct)\big)$, where the distance to the origin r is the parameter of the curve. The resulting solution is a curve rotating around the origin with speed $|c|$. For simplicity, in the following discussion we always take c to be positive so that it rotates in counterclockwise direction. By a direct computation, the curvature of this curve is given by

$$k(r) = \frac{2\alpha'(r) + r\alpha''(r) + r^2\alpha'^3(r)}{(1+r^2\alpha'^2)^{3/2}} ,$$

and the normal velocity is given by

$$\frac{\partial\gamma}{\partial t}\cdot n = \frac{cr}{(1+r^2\alpha'^2)^{3/2}} .$$

Letting $\alpha(r) = \varphi(y)$, $r = y^2$, a solution of the GCSF is a spiral if and only if φ satisfies

$$\frac{d\varphi}{dy} = \frac{\lambda}{2y} , \tag{2.3}$$

$$\frac{d\lambda}{dy} = \frac{1}{2}(1+\lambda^2)\left[cy^{\frac{1}{2\sigma}-\frac{1}{2}}(1+\lambda^2)^{\frac{1}{2}-\frac{1}{2\sigma}} - \frac{\lambda}{y}\right] . \tag{2.4}$$

(2.3) and (2.4) can be combined to yield

$$\left(\tan^{-1}\lambda(y)\right)' = \frac{c}{2}y^{\frac{1}{2\sigma}-\frac{1}{2}}(1+\lambda^2)^{\frac{1}{2}-\frac{1}{2\sigma}} - \varphi'(y) ,$$

that is,

$$\varphi(y) = -\tan^{-1}\lambda(y) + \frac{1}{2}\int^y z^{\frac{1}{2\sigma}-\frac{1}{2}}(1+\lambda(z)^2)^{\frac{1}{2}-\frac{1}{2\sigma}}\,dz . \tag{2.5}$$

In particular, when $\sigma = 1$,

$$\varphi(y) = \frac{c}{2}y - \tan^{-1}\lambda(y) + \text{ constant} .$$

We analyze (2.4) as follows. Setting its right hand side to zero, we have

$$cy^{\frac{1}{2\sigma}+\frac{1}{2}} = \frac{\lambda}{(1+\lambda^2)^{\frac{1}{2}-\frac{1}{2\sigma}}} \ .$$

This equation defines a curve $\Gamma = (y, \Lambda(y))$ in $(0, \infty) \times \mathbb{R}$, where Λ is strictly increasing and satisfies

$$\lim_{y\downarrow 0} \frac{\Lambda(y)}{cy^{\frac{1}{2}+\frac{1}{2\sigma}}} = 1 \ , \tag{2.6}$$

and

$$\lim_{y\longrightarrow\infty} \frac{\Lambda(y)}{c^\sigma y^{\frac{\sigma}{2}+\frac{1}{2}}} = 1 \ . \tag{2.7}$$

The curve Γ divides $(0, \infty) \times \mathbb{R}$ into an upper and a lower region. Any integral curve of (2.4) starting from the upper region strictly decreases as y increases until it hits Λ; then it becomes increasing and tends to Λ asymptotically, never crossing Λ again. On the other hand, any integral curve starting from $(0, \infty) \times (-\infty, 0)$ is strictly increasing and tends to Λ from below as $y \longrightarrow \infty$. Denote by Ω^+ the union of all integral curves starting in the upper region, and by Ω^- the union of all integral curve starting in $(0, \infty) \times (-\infty, 0)$. Both regions are open and Ω^+ lies above Ω^-. Set

$$\lambda^*(y) = \inf\left\{\lambda : (y, \lambda) \in \Omega^+\right\} \ , \text{ and}$$

$$\lambda_*(y) = \sup\left\{\lambda : (y, \lambda) \in \Omega^-\right\} \ .$$

By continuity, both λ^* and λ_* are solutions of (2.4) satisfying $\lambda_*(0) = \lambda^*(0) = 0$. Using (2.4), it is not hard to see that

$$\lambda(y) = \frac{\sigma c}{2\sigma + 1} y^{\frac{1}{2}+\frac{1}{2\sigma}} \left(1 + o(1)\right) \text{ at } y = 0 \ , \tag{2.8}$$

for any solution λ starting at the origin. It follows from (2.8) that the solution satisfies $\lambda(0) = 0$ is = unique. Hence, λ^* and λ_* are identical. From now on we denote it by λ_0. We call a (maximal) solution λ^+ (resp. λ^-) from Ω^+ (resp. Ω^-) a type–I (resp. type–II) solution. It is not hard to see that, for any λ^+ or λ^-, there exists $\delta > 0$ such that

$$\lim_{y \downarrow \delta} \lambda^+(y) = \infty$$

$$\lim_{y \downarrow \delta} \lambda^-(y) = -\infty . \tag{2.9}$$

Moreover, for each $\delta > 0$ there exist a unique type–I and a unique type–II solution which blow up at δ. The graphs of all type–I, type–II solutions, and λ_0 form a foliation of $(0, \infty) \times \mathbb{R}$.

The solution curve $\alpha = \varphi(y) + ct$ depends on two parameters because φ satisfies a second order ODE. One of the parameters is the choice of λ and the other is the integration constant arising from integrating (2.3). The latter accounts for a rotation of the solution curve in the (x, u)-plane and is not essential. We shall always assume that the solution curve starts at the positive x-axis. The solution curve rotates in unit speed with its shape unchanged. It suffices to look at the snapshot at $t = 0$, i.e., $\varphi(y)$.

First of all, let φ be a solution of (2.3) where $\lambda \in (\delta, \infty)$, $\delta \geqslant 0$. Its curvature is given by

$$k(r) = \frac{cr^{\frac{1}{\sigma}}}{(1 + \lambda^2(r))^{\frac{1}{2\sigma}}} . \tag{2.10}$$

Hence, $k(r) \longrightarrow 0$ as $r \downarrow \delta$ and $k(r) = O\left(\frac{c}{r}\right)$ at $r = \infty$. We also have

$$\alpha(r) = \frac{cr^{1+\sigma}}{1+\sigma}(1 + o(1)) \quad \text{at} \quad r = \infty .$$

When λ is equal to λ_0 or of type–I, α increases from 0 to ∞ as y increases from δ to ∞. When λ is of type–II, there exists $\delta' > \delta$ such that $\lambda(\delta') = 0$. So α decreases for $y \in (\delta, \delta')$ and increases to ∞ in

(δ', ∞).

The regularity of the spiral with its tip at the origin can be read from (2.10). As $ds/dr = (1 + \lambda^2)^{1/2}$, where ds is the arc-length element, we see that the spiral is smooth at the tip if and only if $1/\sigma$ is a non-negative integer. When $1/\sigma$ is odd, we can extend it by setting

$$\big(x(r), y(r)\big) = -\big(x(-r), y(-r)\big) \,, \quad r \in (-\infty, 0) \,.$$

The resulting curve is a smooth, complete, simple curve whose curvature is positive for all non-zero r. It is called the "yin-yang" curve when $\sigma = 1$. It is possible to match a type–I solution with a type–II solution to form a smooth, complete, embedded asymmetric spiral. To see this, let's choose λ^+ and λ^- so that they blow up at the same δ. The corresponding curves $\gamma^\pm$ start at $(0, \sqrt{\delta})$. We joint γ^- to γ^+ to form a complete, embedded γ. We claim that it is smooth at $(0, \sqrt{\delta})$. For, since $d\alpha/dr > 0$ on γ^+ and $d\alpha/dr < 0$ on γ^- near $(0, \sqrt{\delta})$, we may represent γ as a function $y = y(\alpha)$, for $\alpha \in (-\varepsilon, \varepsilon)$, ε small. When $\alpha \in (-\varepsilon, 0]$, (resp. $\alpha \in [0, \varepsilon)$), γ is γ^+ (resp. γ^-). To show that y is smooth at 0 we write (2.4) as

$$-L_\theta = (1 + L^2)\Big[cy^{\frac{1}{2\sigma}-\frac{1}{2}}(1 + L^2)^{\frac{1}{2}-\frac{1}{2\sigma}}L^{\frac{1}{\sigma}}y - 1\Big] \,, \qquad (2.11)$$

where $L = 1/\lambda$ is regarded as a function of α. Notice that L is continuous in $(-\varepsilon, \varepsilon)$ and $L(0) = 0$. On the other hand, from (2.3) we have

$$y_\theta = 2yL \,. \qquad (2.12)$$

(2.11) and (2.12) together form a system of ODEs for y and L. Since the pair (y, L) is continuous in $(-\varepsilon, \varepsilon)$, it is also smooth there.

In conclusion, we have proved the following results. *Given any positive c and point X in $\mathbb{R}^2$, there exists a spiral of the GCSF rotating around the origin in constant speed c in the counterclockwise* *direction. Moreover, it has a unique inflection point at X. The spiral is unique when $X \neq (0,0)$ and unique up to a rotation when $X = (0,0)$.*

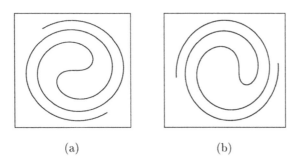

(a)   (b)

Figure 2.1

Spirals for $\sigma = 1$: (a) the yin-yang curve and (b) an asymmetric spiral.

2.3 The support function of a convex curve

Before we discuss self-similar solutions of the generalized curve shortening flow, we make a digression to review some basic facts of the support function for a convex curve.

Let γ be a curve with positive curvature, i.e, it is **uniformly convex**. The **normal angle**[1] θ at a point $\gamma(p)$ is defined by $n(p) = -(\cos\theta, \sin\theta)$. It is determined modulo 2π. Just as the tangent angle, once we fix its value at a certain point on the curve $\gamma(p)$, the normal angles at other points on the curve are fixed by continuity. The image of I under the normal map $\theta(p)$, J, is an open interval (θ_1, θ_2) and γ is closed only if $\theta_2 - \theta_1$ is a multiple of 2π. We may use the normal angle to parametrize the curve. In the following, we let $\gamma = \gamma(\theta)$ so that $n = -(\cos\theta, \sin\theta)$. The **support function** of γ is a function defined in J given by $h(\theta) = \langle \gamma(\theta), (\cos\theta, \sin\theta) \rangle$. We

---

[1]The Greek letter θ is used to denote the tangent or the normal angle in different context. In the definitions for the support function, equations (3.9) and (4.1), it stands for the normal angle, while it means the tangent angles in other places.

have

$$h_\theta(\theta) = -\gamma^1 \sin\theta + \gamma^2 \cos\theta \ .$$

Therefore, γ can be recovered from h by

$$\begin{cases} \gamma^1 = h\cos\theta - h_\theta \sin\theta, \\[2mm] \gamma^2 = h\sin\theta + h_\theta \cos\theta \ . \end{cases} \tag{2.13}$$

The curvature of γ can be expressed in terms of the support function in a very neat form. In fact,

$$\begin{aligned} h_{\theta\theta} + h \ &= \ -\gamma^1_\theta \sin\theta + \gamma^2_\theta \cos\theta \\[2mm] &= \ \left\langle \frac{d\gamma}{ds}\frac{ds}{d\theta}, \ t \right\rangle \\[2mm] &= \ \frac{ds}{d\theta} \ . \end{aligned}$$

By the Frenet formulas, we have

$$h_{\theta\theta} + h = \frac{1}{k} \ . \tag{2.14}$$

The support function can be described as follows. Let ℓ be the tangent line passing a point P on the curve whose normal is $-(\cos\theta, \sin\theta)$. Then, the support function is the signed distance from the origin O to ℓ; it is positive (resp. negative) if the angle between $\overline{OP}$ and $(\cos\theta, \sin\theta)$ is acute (resp. obtuse). We note that the support function depends on the choice of the origin. When O is changed to O', the support function is changed from h to $h - \langle \overline{OO'}, (\cos\theta, \sin\theta) \rangle$.

The relationship between uniformly convex, closed curves and their support functions is contained in the following proposition whose proof is straightforward.

Proposition 2.1 *Any $2n\pi$-periodic function h with $h_{\theta\theta} + h > 0$ determines a closed, uniformly convex curve by (2.13) whose support function is u. Two curves determined by h and h', respectively, differ by a translation if and only if the difference of h and h' is equal to $C_1 \cos\theta + C_2 \sin\theta$ for some C_1 and C_2. Moreover, γ is embedded if and only if h is 2π-periodic.*

Let $\gamma(\cdot, t)$ be a family of uniformly convex curves satisfying (1.2). We have

$$\frac{\partial \tilde{\gamma}}{\partial t} = \gamma_p \frac{\partial p}{\partial t} + F\boldsymbol{n} \ ,$$

where $\tilde{\gamma}(\theta, t) = \gamma(p(\theta, t), t)$. Consequently, $h(\theta, t) = \langle \tilde{\gamma}(\theta, t), -\boldsymbol{n} \rangle$ satisfies

$$h_t = -F(\gamma, \theta + \frac{\pi}{2}, k) \ , \tag{2.15}$$

where γ and k are given in (2.13) and (2.14), respectively, and we have chosen the tangent angle to be $\theta + \pi/2$. It is a parabolic equation if F is parabolic. For convex flows, this is the most convenient way to reduce (1.2) to a single equation. Notice again by Proposition 1.1, (2.15) is equivalent to (1.2) as long as $\gamma(\cdot, t)$ is closed and uniformly convex.

2.4   Self-similar solutions

Now let's return to the GCSF. We seek solutions of this flow whose shapes change homothetically during the evolution: $\hat{\gamma}(\cdot, t) = \lambda(t)\gamma(\cdot)$. According to Proposition 1.1, $\hat{\gamma}$ is a self-similar solution if and only if its normal velocity is equal to $|\hat{k}|^{\sigma-1}\,\hat{k}$. We have

$$\lambda'\lambda^\sigma \gamma \cdot \boldsymbol{n} = |k|^{\sigma-1}k.$$

When this curve is not flat, $\lambda'\lambda^\sigma$ must be a non-zero constant. After a rescaling, we may simply assume the constant is 1 or -1. When it is

1, $\lambda(t) = [(1+\sigma)t + \text{const.}]^{\frac{1}{1+\sigma}}$ and so $\widehat{\gamma}$ expands as t increases. When it is -1, $\lambda(t) = [-(1+\sigma)t + \text{const.}]^{\frac{1}{1+\sigma}}$, and so $\widehat{\gamma}$ contracts as to a point t increases up to some fixed time. We call the former an **expanding self-similar solution** and the latter a **contracting self-similar solution**. Noticing that $k\mathbf{n}$ is independent of the orientation of the curve, we may assume that k is positive somewhere. By introducing the normal angle and the support function h, it is readily seen that h satisfies the following ODE,

$$h_{\theta\theta} + h = \frac{1}{(-h)^p} \ , \quad \text{(expanding self-similar curve)} \qquad (2.16)$$

or

$$h_{\theta\theta} + h = \frac{1}{h^p} \ , \quad \text{(contracting self-similar curve)} \ , \qquad (2.17)$$

where $p = 1/\sigma$.

We shall study (2.16) and (2.17) for $\sigma \leqslant 1$. First, let's denote the solution of (2.16) subject to the initial conditions $h(0) = -\alpha, \alpha > 0$ and $h_\theta(0) = 0$ by $h(\theta, \alpha)$. Clearly, for each $\alpha \in (0, \infty)$, $h(\cdot, \alpha)$ is an even, convex function which is strictly increasing in $(0, \theta(\alpha))$ where $\theta(\alpha)$ is the zero of $h(\cdot, \alpha)$ and $h_\theta(\cdot, \alpha)$ blows up as $\theta \uparrow \theta(\alpha)$. Further, we claim that $h(\cdot, \alpha_1) > h(\cdot, \alpha_2)$ for $\alpha_1 < \alpha_2$ and $\theta(\alpha)$ is strictly increasing from 0 to $\pi/2$ as α goes from 0 to ∞. To see this, we first note that $w = h(\cdot, \alpha_1) - h(\cdot, \alpha_2)$ satisfies $w_{\theta\theta} + (1 + q(\theta))w = 0$ where q is negative. By the Sturm comparison theorem $h(\cdot, \alpha_1) > h(\cdot, \alpha_2)$ in $(-\pi/2, \pi/2)$. Next, the "normalized" support function $\widehat{h} = h/\alpha$ satisfies $\widehat{h}(0) = 1$, $\widehat{h}_\theta(0) = 0$ and it converges to $\cos\theta$ on every compact subset of $(-\pi/2, \pi/2)$ as $\alpha \to \infty$. Hence, the claims hold. Consequently, for any Θ in $(0, \pi/2)$, the function $h(\theta + 3\pi/2, \alpha)$, where α is given by $\Theta = \theta(\alpha)$, determines a complete expanding self-similar solution $\Gamma(\Theta)$ lying inside the sector $\{(x, y) : y < |x|\tan\Theta\}$; it is the graph of a convex function u satisfying $u(x) \mp (\tan\Theta)x \to 0$ and $u_x \mp \tan\Theta \to 0$ as $x \to \pm\infty$. In summary, we have

Proposition 2.2 *Every expanding self-similar solution is contained in a complete expanding self-similar solution, which is congruent to exactly one $\Gamma(\Theta)$ after the multiplication of a constant.*

Next, denote the solution of (2.17) subject to the initial conditions $h(0) = \alpha \geqslant 1$, $h_\theta(0) = 0$, by $h(\theta, \alpha)$. We note that it admits a first integral

$$\frac{1}{2}(h_\theta^2 + h^2) - \frac{h^{1-p}}{1-p}, \qquad (p \neq 1),$$

or

$$\frac{1}{2}(h_\theta^2 + h^2) - \log h, \qquad (p = 1).$$

It follows that, for each α, the function $h(\cdot, \alpha)$ is periodic in θ. Its maximum $M = h(0)$ and minimum m are related by

$$\frac{1}{2}(M^2 - m^2) = \frac{1}{1-p}(M^{1-p} - m^{1-p}), \qquad (p \neq 1)$$

or

$$\frac{1}{2}(M^2 - m^2) = \log \frac{M}{m}, \qquad (p = 1).$$

Hence, when $p \geqslant 1$, $m \to 0$ as $M = \alpha \to \infty$. Also, it is clear that h is concave on $\{h > 1\}$ and convex on $\{h < 1\}$. We claim that, for any $T \in (\pi, 2\pi/\sqrt{p+1})(p < 3)$ or $(2\pi/\sqrt{p+1}, \pi)(p > 3)$, there exists some $\alpha > 1$ such that $h(\cdot, \alpha)$ has period T. To see this, we first observe that $h(\cdot, 1) \equiv 1$ is a trivial solution, which corresponds to the unit circle. The linearized equation of (2.17) at this trivial solution is given by

$$\phi'' + (1+p)\phi = 0, \qquad \phi'(0) = 0 .$$

It admits $\varphi = \cos \sqrt{p+1}\,\theta$ as its solution. Therefore, for small ε, $h(\theta, 1+\varepsilon) = 1 + \varepsilon \cos \sqrt{p+1}\,\theta + O(\varepsilon^2)$, and so the period of $h(\cdot, 1+\varepsilon)$ is very close to $2\pi/\sqrt{p+1}$. On the other hand, it is not hard to see that the normalized support function h/α tends to $|\cos \theta|$ uniformly on every compact set in $\mathbb{R}\backslash\{\pi/2 + n\pi,\ n \in \mathbb{Z}\}$. By continuity, we conclude that every number between π and $2\pi/\sqrt{p+1}$ is a period of h. When $p = 3$, all solutions of (2.17) are given by

$$(\alpha^2 \cos^2(\theta - \theta_0) + \alpha^{-2} \sin^2(\theta - \theta_0))^{1/2}, \quad \alpha \in [1, \infty), \theta_0 \in [0, \pi) \ .$$

They are support functions of ellipses.

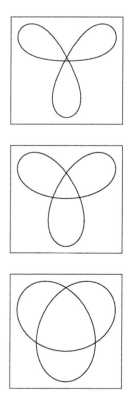

Figure 2.2

Contracting self-similar 3-petal curves for $\sigma = 0.7, 1, 1.2$.

We shall use the following characterization of the circle in the next chapter.

Proposition 2.3 *For $\sigma > 1/3$, the only closed, embedded contracting self-similar solutions are circles.*

Proof: Let $h = h(\theta, \alpha)$ be a 2π-periodic solution of (2.17) satisfying $h(0) = \alpha > 1$ and $h_\theta(0) = 0$. We have

$$(h_\theta)_{\theta\theta} + \left(1 + \frac{p}{h^{p+1}}\right) h_\theta = 0.$$

By the Sturm comparison theorem, h_θ must vanish at some point in $(0, \pi)$. But then it means that h is symmetric with respect to the point. So the minimal period of h, T, is at most π.

Next, we have

$$\frac{1}{2}\left[(h^2)_{\theta\theta\theta} + 4(h^2)_\theta\right] = \frac{(3-p)h_\theta}{h}.$$

Therefore,

$$2(3-p)\int_0^T \frac{h_\theta}{h}\cos 2(\theta - \theta_0)d\theta = \int_0^T \cos 2(\theta - \theta_0)[(h^2)_{\theta\theta\theta} + 4(h^2)_\theta]$$

$$= \cos 2(\theta - \theta_0)(h^2)_{\theta\theta}\Big|_0^T.$$

If we choose $2\theta_0 = (T - \pi/2)$, the right hand side of this identity becomes

$$-4\left(\frac{1}{h^{p-1}}(0) - h^2(0)\right)\cos 2\theta_0,$$

which is positive because $T \leqslant \pi$. On the other hand, $\cos 2(T/2 - \theta_0) = \cos \pi/2 = 0$ and $h_\theta < 0$ (resp. > 0) in $(0, T/2)$ (resp. $(T/2, T)$). Therefore, its left-hand side is negative. The contradiction holds. We have shown that the only 2π-periodic solution of (2.17) is $h(\cdot, 1)$. $\square$

In the rest of this section, we examine closed contracting self-similar solutions for the CSF more closely. They are sometimes called **Abresch-Langer curves**.

It was shown in Abresch-Langer [1] that the period of k, $T(\alpha)$, strictly decreases from $\sqrt{2}\pi$ to π as α increases from 1 to ∞. Consequently, for any pair of relatively prime natural numbers m and n, satisfying $m/n \in (1, \sqrt{2})$, there corresponds a closed Abresch-Langer curve which rotates around the origin m times and has n petals. Moreover, no other values in $(1, \sqrt{2})$ yield closed curves. To reconcile the notations here with those in [1], we note that the first integral of (2.17) ($p = 1$) is

$$\frac{1}{2}h_\theta^2 + \frac{1}{2}h^2 = \log h + \eta + \frac{1}{2} \, ,$$

where $\eta = \frac{1}{2}\alpha^2 - \frac{1}{2} - \log \alpha$ satisfies $\eta(1) = 0$, i.e., it vanishes at the unit circle. On the other hand, letting $B = 2 \log k$ and observing that $k = h$, it follows from (2.17) that

$$B_{ss} + 2(e^B - 1) = 0 \, ,$$

and

$$\frac{1}{2}B_s^2 + 2(e^B - B - 1) = 2\eta \, .$$

They are precisely equations (2) and (3) in [1] where we have taken λ to 1. The total change in tangent\normal angle within a period of k is given by

$$\begin{aligned}
\Theta(\eta) &= \int_{B_{\min}}^{B_{\max}} \frac{2dB}{\sqrt{4e^{-B}[\eta - (e^B - B - 1)]}} \\
&= \int_{h_{\min}}^{h_{\max}} \frac{2dh}{\sqrt{2\eta - (h^2 - 2\log h - 1)}} \\
&= T(\alpha) \, .
\end{aligned} \qquad (2.18)$$

Proposition 3.2 (v) in [1] asserts that $\Theta(\eta)$ is strictly decreasing in η, and so T is strictly decreasing in α.

Next, we consider the linearized stability of the Abresch-Langer curve. Let $w_{m,n}$ be the support function of the Abresch-Langer curve corresponding to (m, n). Consider the eigenvalue problem with periodic boundary condition in the space S_m consisting of all periodic functions in $[0, 2m\pi]$,

$$\mathcal{L}\varphi = \varphi_{\theta\theta} + \left(1 + \frac{1}{w^2}\right)\varphi$$
$$= \frac{-\lambda}{w^2}\varphi, \tag{2.19}$$

where $w = w_{m,n}$. It is well-known that (2.19) has infinitely many eigenvalues satisfying

$$\lambda_0 < \lambda_1 \leqslant \lambda_2 < \lambda_3 \leqslant \lambda_4 < \cdots \longrightarrow \infty,$$

and the corresponding eigenfunctions φ_{2j-1} and φ_{2j} have exactly $2j$ zeroes in S_m. One can verify that $\varphi_0 = w$ and $\lambda_0 = -2$. Also, the functions $\cos\theta$ and $\sin\theta$ are eigenfunctions for the eigenvalue -1.

Proposition 2.4 *When $w \not\equiv 1$, w_θ is the $(2n-1)-st$ eigenfunction of (2.19) and $\lambda_{2n-1} = 0$. Moreover, zero is a simple eigenvalue.*

Proof: Since $w \not\equiv 1$, w_θ is nontrivial and satisfies (2.19) for $\lambda = 0$. As it has exactly $2n$ zeroes, it is either the $(2n-1)-st$ or the $2n-th$ eigenfunction. To show that it is the former, we consider the Neumann eigenvalue problem in $[0, m\pi/n]$:

$$\begin{cases} \mathcal{L}\varphi = \dfrac{-\lambda_N}{w^2}\varphi, \\[2mm] \varphi_\theta(0) = \varphi_\theta\left(\dfrac{m\pi}{n}\right) = 0. \end{cases} \tag{2.20}$$

Since w is even, each eigenfunction of (2.20) can be extended by odd function and then periodically to an eigenfunction of (2.19) with the same eigenvalue. In particular, the eigenfunction φ_2 corresponding to λ_N^2 has exactly $2n$ zeroes in S_m. As w_θ is the lowest Dirichlet eigenvalue of $\mathcal{L}$ in $[0, m\pi/2]$ and it is well-known that $\lambda_1^D > \lambda_2^N$, w_θ

and φ_2 are, respectively, the $(2n-1)-st$ and $2n-th$ eigenfunctions of (2.19).

To show that zero is simple, we recall that $w(\theta) = h(\theta, \alpha_0)$ for some $\alpha_0 > 1$. So $u(\theta) = \partial h(\theta, \alpha)/\partial\alpha\big|_{\alpha=\alpha_0}$ is also a solution of (2.19) with $\lambda = 0$. However, since the period strictly decreases in α, u cannot belong to S_m. So zero must be a simple eigenvalue.   □

It follows that, when $w \not\equiv 1$, the stable invariant manifold of w is of codimension $2n$ and the unstable invariant manifold has dimension $2n-1$. When $w \equiv 1$, the eigenfunctions of (2.19) can be found explicitly. It is straightforward to verify that the stable invariant manifold of w is of codimension N_+ and the unstable invariant manifold has dimension N_+ where

$$N_+ = \left|\left\{q \in \mathbb{Z} : 2 - \left(\frac{q}{m}\right)^2 > 0\right\}\right|.$$

Notice that now zero is no longer an eigenvalue of (2.19).

Notes

The grim reapers were used in Grayson [66], Altschuler [3], and Hamilton [72]. It can be characterized as the eternal solution (a solution exists for all $t \in \mathbb{R}$) with non-negative curvature. Its stability was studied in Altschuler-Wu [5]. In the context of the CSF, the spiral was first pointed out in Altschuler [3], where it is called the "Yin-Yang curve." However, spirals arising from the curvature-eikonal flow have been studied extensively in mathematical biology and chemical reaction. See, for instance, Keener-Sneyd [84], Murray [90], and Ikota-Ishimura-Yamaguchi [80]. The expanding self-similar solution for the mean curvature flow was known back to Brakke [22], who called it "evolution from a corner." See Ecker-Huisken [46] for a result concerning its stability. The Abresch-Langer curves were classified in [1] and further results on their stability can be found in this paper and Epstein-Weinstein [49]. Proposition 2.4 is taken from [49]. Our discussion on the linearized stability slightly differs from [49] because they use the equation for $k = k(s)$ instead of (2.17). Notice that, in their formulation, the eigenfunctions $\cos \theta$ and $\sin \theta$, which correspond to translating the curve in the plane, do not come up. Finally, we mention that a certain contracting spiral was used in the study of the formation of singularities in Angenent [15] and Angenent-Velazquez [18].

All special solutions discussed in this chapter are group invariant solutions. A systematic investigation of the group invariant solutions for the GCSF can be found in Chou-Li [31]. Using Lie's theory of symmetries for differential equations, they determine the symmetry group and obtain the optimal system (the largest family consisting of mutually nonequivalent group invariant solutions) for the GCSF. Many new special solutions are found for the affine CSF. In this connection, also see Calabi-Olver-Tannenbaum [25].

Chapter 3

The Curvature-Eikonal Flow for Convex Curves

In this chapter, we shall study the anisotropic curvature-eikonal flow (ACEF) for closed convex curves. Three cases—contracting to a point, converging to a stationary shape, and expanding to infinity—will be studied in detail in Sections 4, 5 and 6. In Section 3, we present a self-contained treatment of an important special case—the curve shortening flow. Some sufficient conditions which ensure shrinking to a point are established in Section 2. They apply to flows more general than the ACEF and will be used in later chapters.

3.1 Blaschke Selection Theorem

The following basic compactness result for convex sets is very basic.

Theorem 3.1 (Blaschke Selection Theorem) *Let $\{K_j\}$ be a sequence of convex sets which are contained in a bounded set. Then there exists a subsequence $\{K_{j_k}\}$ and a convex set K such that K_{j_k} converges to K in the Hausdorff metric.*

Recall that the Hausdorff metric for two arbitrary sets A and B are given by

$$d(A, B) = \inf \left\{ \lambda \geqslant 0 \; : \; A \subseteq B + \lambda D_1 \;,\; B \subseteq A + \lambda D_1 \right\},$$

where D_1 is the unit disk centered at the origin.

The convergence in Hausdorff metric for convex sets is related to the uniform convergence of their support functions. In fact, for any convex set K, its support function is given by

$$h(\theta) = \sup \left\{ \langle (x, y), (\cos \theta, \sin \theta) \rangle \; : \; (x, y) \in K \right\}.$$

When the boundary of K, ∂K, is uniformly convex, this definition is the same as the one given in §2.3. We extend h to become a function of homogeneous degree one in the whole plane by setting

$$H(x, y) = rh(\theta) \quad , \quad (x, y) = (r \cos \theta, r \sin \theta).$$

Then H is convex. To see this, we note that, when ∂K is uniformly convex, the Hessian matrix of H is given by

$$\frac{h_{\theta\theta} + h}{r} \begin{bmatrix} \sin^2 \theta & -\sin \theta \cos \theta \\ \\ -\sin \theta \cos \theta & \cos^2 \theta \end{bmatrix},$$

and it is non-negative definite. The general case then follows by approximation.

There is a one-to-one correspondence between bounded convex sets and convex functions of homogeneous degree one in the plane. Actually, any such function H determines a convex set defined by

$$K = \left\{ (x, y) : \langle (x, y), (\cos \theta, \sin \theta) \rangle \leqslant h(\theta), \text{ for all } \theta \in [0, 2\pi) \right\}.$$

It is not hard to check that the support function of K is h.

Now, it is clear that $\{K_j\}$ converges to K in the Hausdorff metric if and only if $\{H_j\}$ converges to H uniformly in every compact set. We can see why the Blaschke Selection Theorem holds using this connection. First of all, by convexity,

$$\sup_{C_1} |\nabla H| \leqslant \frac{\sup_{C_2} |H|}{\mathrm{dist}\,(C_1, \partial C_2)}$$

for any bounded open sets C_1 and C_2 with $\overline{C_1} \subseteq C_2$. When $\{K_j\}$ is confined to a bounded set, $\{H_j\}$ is uniformly bounded. Hence, we can select a subsequence $\{H_{j_k}\}$ which converges uniformly to some convex function H. In other words, $\{K_{j_k}\}$ converges to the convex set determined by H in the Hausdorff metric.

3.2 Preserving convexity and shrinking to a point

We first show that convexity is preserved for a large class of (1.2).

Proposition 3.2 (preserving convexity) *Let γ be a maximal solution of (1.2) in $C^2\left(S^1 \times [0, \omega)\right)$. Then, $\gamma(\cdot, t)$ is uniformly convex for positive t under either one of the following conditions:*

(a) F is parabolic in $\mathbb{R}^2 \times S^1 \times (0, \infty)$ and γ_0 is uniformly convex ;

(b) F is parabolic in $\mathbb{R}^2 \times S^1 \times [0, \infty)$, $\partial^2 F/\partial x^2 = \partial^2 F/\partial x \partial y = \partial^2 F/\partial y^2 = 0$ at $q = 0$, and γ_0 is convex.

Proof: We first prove the proposition under (a). Introduce the support function as long as $\gamma(\cdot, t)$ is uniformly convex. By (2.14) and (2.15), the curvature $k = k(\theta, t)$ satisfies

$$\frac{\partial k}{\partial t} = k^2 \left(\frac{\partial^2 F}{\partial \theta^2} + F \right) . \tag{3.1}$$

We may rewrite the equation as

$$k_t = k^2 F_q k_{\theta\theta} + Gk ,$$

where G is bounded in $S^1 \times [0,T]$ for any $T < \omega$. The function $v = e^{\lambda t}k$ satisfies

$$v_t = k^2 F_q v_{\theta\theta} + (G + \lambda)v .$$

We choose λ so that $G+\lambda$ is non-negative. Let $v_{\min}(t) = \min\{v(\theta,t) : \theta \in [0,2\pi)\}$. It is not hard to show that $v_{\min}$ in Lipschitz continuous and $dv_{\min}(t)/dt \geqslant \partial v/\partial t(\theta,t)$ at $v_{min}(t) = v(\theta,t)$. So,

$$\frac{dv_{\min}}{dt} \geqslant (G+\lambda)v_{\min} \geqslant 0 .$$

In other words, $k(\theta,t) \geqslant e^{-\lambda t}\min\limits_{\theta} k(\theta,0) > 0$.

Next, assume (b) holds. It suffices to show that k becomes positive instantly. Let's look at the local graph representation of the flow. By differentiating (1.3) and using (b) $k = k(x,t)$ satisfies a parabolic equation of the form $k_t = (1 + u_x^2)^{-\frac{1}{2}}F_q k_{xx} + bk_x + ck$, where b and c are bounded. It follows from the strong maximum principle that k is positive for $t > 0$. $\square$

Proposition 3.3 Let $\gamma(\cdot,t)$ be a uniformly convex solution of (1.2) where F is parabolic in $\mathbb{R}^2 \times S^1 \times (0,\infty)$. Suppose further that F satisfies (i) there exists a constant C_0 such that $F(x,y,\theta,0) \geqslant -C_0$ in $\mathbb{R}^2 \times S^1$, (ii) for each (x,y,θ), $\lim\limits_{q\to\infty} F(x,y,\theta,q) = \infty$, and (iii) for any compact set K in $\mathbb{R}^2 \times S^1$, there exists a constant C_1 such that

$$|F| + |F_x| + |F_y| \leqslant C_1(1 + qF_q)$$

in $\mathbb{R}^2 \times S^1 \times [0,\infty)$. Then, if ω is finite, the area enclosed by $\gamma(\cdot,t)$ tends to zero as $t \uparrow \omega$.

Proof: First of all, by parabolicity, assumption (i) and Proposition 1.6, $\gamma(\cdot,t)$ are confined to a disk of radius $diam\ \gamma_0 + C_0\omega$. Suppose on the contrary that there exists a sequence $\{t_j\}$, $t_j \uparrow \omega$ such that the area enclosed by $\gamma(\cdot,t_j)$ does not tend to zero. By the Blaschke

Selection Theorem, we may select a subsequence, still denoted by $\{\gamma(\cdot, t_j)\}$, such that each $\gamma(\cdot, t_j)$ contains a disk D_j, and $\{D_j\}$ converges to the disk $D_{4\rho}((x_0, y_0))$ for some (x_0, y_0) and ρ. By applying the strong separation principle to $\gamma(\cdot, t)$ and the flow starting at $\partial D_{4\rho}((x_0, y_0))$, we know that $D_{2\rho}((x_0, y_0))$ is enclosed by $\gamma(\cdot, t)$ for all t sufficiently close to ω.

Use (x_0, y_0) as the origin and introduce the support function $h(\theta, t)$. Then $h(\theta, t) \geq 2\rho$ on $[t_0, \omega)$ where t_0 is close to ω. We claim that the curvature is uniformly bounded in $[t_0, \omega)$. To prove this, we consider the function

$$\Phi = \frac{-h_t}{h - \rho} = \frac{F}{h - \rho}.$$

Let $\Phi(\theta_0, t_1) = \max\{\Phi(\theta, t) : (\theta, t) \in S^1 \times [t_0, T]\}$ where $T < \omega$. In view of (ii), we may assume $\Phi(\theta_0, t_1)$ is positive. If $t_1 > t_0$, we have

$$0 = \Phi_\theta = \frac{-h_{\theta t}}{h - \rho} + \frac{h_t h_\theta}{(h - \rho)^2}, \tag{3.2}$$

$$0 \leq \Phi_t = \frac{-h_{tt}}{h - \rho} + \frac{h_t^2}{(h - \rho)^2}, \quad \text{and} \tag{3.3}$$

$$0 \geq \Phi_{\theta\theta} = \frac{-h_{\theta\theta t}}{h - \rho} + \frac{2h_{\theta t} h_\theta}{(h - \rho)^2} + \frac{h_t h_{\theta\theta}}{(h - \rho)^2} - \frac{2h_t h_\theta^2}{(h - \rho)^3}, \tag{3.4}$$

at (θ_0, t_1). On the other hand, from (2.15) and (3.1), we have

$$h_{tt} = -F_x \frac{\partial \gamma^1}{\partial t} - F_y \frac{\partial \gamma^2}{\partial t} - F_q k^2 \left(\frac{\partial^2 F}{\partial \theta^2} + F\right).$$

Therefore, by (3.3), (3.2), and (3.4),

$$0 \leq -F_x \frac{\partial \gamma^1}{\partial t} - F_y \frac{\partial \gamma^2}{\partial t} - F_q k^2 \left(-h_{\theta\theta t} + F\right) + \frac{F^2}{h - \rho}$$

$$\leq -F_x \frac{\partial \gamma^1}{\partial t} - F_y \frac{\partial \gamma^2}{\partial t} + \frac{F F_q k}{h - \rho} - \frac{\rho F F_q k^2}{h - \rho} + \frac{F^2}{h - \rho}.$$

By (2.13), we have

$$F_x \frac{\partial \gamma^1}{\partial t} + F_y \frac{\partial \gamma^2}{\partial t} = -F\left(F_x \cos\theta + F_y \sin\theta\right) + \frac{F(F_x \sin\theta) - F_y(\cos\theta)h_\theta}{h - \rho} \; .$$

By convexity, we know that $|h_\theta| \leqslant \pi \sup |h|$. Therefore, it follows from (iii) that $\Phi(\theta_0, t_1)$ is bounded above. But then (ii) implies that the curvature is uniformly bounded in $[t_0, \omega)$, which is impossible by Proposition 1.2. Hence, the enclosed area must tend to zero.   □

Proposition 3.4 (shrinking to a point) *Let $\gamma(\cdot, t)$ be a uniformly convex solution of (1.2) where F is parabolic in $\mathbb{R}^2 \times S^1 \times (0, \infty)$. Suppose that ω is finite and the enclosed area tends to zero as $t \uparrow \omega$. Then, $\gamma(\cdot, t)$ shrinks to a point under either one of the following conditions:*

(i) $F(x, y, \theta, 0) = 0$, or

(ii) F is uniformly parabolic in $\mathbb{R}^2 \times S^1 \times [0, \omega)$ and $F(x, y, \theta, 0)$ only depends on θ.

Proof: By (3.1), we have

$$\frac{\partial F}{\partial t} = k^2 F_q \left(\frac{\partial^2 F}{\partial \theta^2} + F \right) \; .$$

So

$$\frac{dF_{\min}}{dt} \geqslant k^2 F_q F_{\min} \; .$$

In case the flow does not tend to a point as the enclosed area approaches to zero, clearly, $k_{\min}(t)$ tends to zero. According to (i), $F_{\min}(t)$ tends to zero too. But, the inequality above shows that $F_{\min}$ has a positive lower bound. So the flow must shrink to a point.

Next, assume (ii) is valid. By uniform parabolicity, $F \leqslant C(1+k)$. By comparing the flow with a large expanding circle with normal velocity $-C(1 + k)$, we know that γ is contained in a bounded set. If it does not shrink to a point, we can find $\{\gamma(\cdot,t_j)\}$, $t_j \uparrow \omega$, such that $\gamma(\cdot,t)$ converges to a line segment. For simplicity we may take this segment to be $\{(x,0) : |x| \leqslant \ell/2\}$. By (ii), vertical lines translate in constant speed under the flow. By the Sturm Oscillation Theorem, for all t sufficiently close to ω, $\gamma(\cdot,t)$ over $(-\ell/4, \ell/4)$ is the union of the graphs of two functions U_1 and U_2, $U_1(x,t) < U_2(x,t)$, which converge to 0 together with bounded gradients. Now we can follow the argument in the proof of Proposition 1.9 that the curvature of γ over $[-\ell/8, \ell/8]$ is uniformly bounded near ω. However, then by the strong maximum principle, the curvature cannot become zero at $t = \omega$. The contradiction holds. $\square$

3.3   Gage-Hamilton Theorem

The main result of the curve shortening flow for convex curves is contained in the following two theorems of Gage and Hamilton.

Theorem 3.5 *Consider the Cauchy problem for the curve shortening flow, where γ_0 is a convex, embedded closed curve. It has a unique solution $\gamma(\cdot,t)$ which is analytic and uniformly convex for each t in $(0,\omega)$ where $\omega = A_0/2\pi$ and A_0 is the area enclosed by γ_0. As $t \uparrow \omega$, $\gamma(\cdot,t)$ shrinks to a point.*

This theorem follows from Proposition 1.1, 3.2–3.4, and (1.17).

To study the longtime behavior of the flow, we rescale the curve by setting

$$\widetilde{\gamma}(\cdot,t) = \left(\frac{\pi}{A(t)}\right)^{\frac{1}{2}} \left(\gamma(\cdot,t) - \gamma(\cdot,\omega)\right),$$

where $A(t) = A_0 - 2\pi t$. Then the area enclosed by this normalized curve $\widetilde{\gamma}$ is always equal to π. The normalized support function and curvature are given by

$$\tilde{h}(\cdot, t) = \left(\frac{\pi}{A(t)}\right)^{1/2} h(\cdot, t) \ ,$$

and

$$\tilde{k}(\cdot, t) = \left(\frac{A(t)}{\pi}\right)^{1/2} k(\cdot, t) \ ,$$

respectively. They satisfy the equations

$$\tilde{h}_\tau = -\tilde{k} + \tilde{h} \ , \tag{3.5}$$

and

$$\tilde{k}_\tau = \tilde{k}^2 \left(\tilde{k}_{\theta\theta} + \tilde{k}\right) - \tilde{k} \ , \tag{3.6}$$

where $2\tau = -\log(A_0 - 2\pi t)$. The normalized flow $\tilde{\gamma}(\cdot, \tau)$ is defined in $[\tau_0, \infty)$ for $\tau_0 = -2^{-1} \log A_0$.

Theorem 3.6 $\tilde{\gamma}$ *converges to the unit circle centered at the origin smoothly and exponentially as* $\tau \longrightarrow \infty$.

The proof of this theorem will be accomplished in several steps. First we have the following gradient estimate.

Lemma 3.7

$$\sup_{S^1 \times [\tau_0, \tau]} \left(\tilde{k}_\theta^2 + \tilde{k}^2\right) \leqslant \max\left\{ \sup_{S^1 \times [\tau_0, \tau]} \tilde{k}^2 \ , \ \sup_{S^1 \times \{\tau_0\}} \left(\tilde{k}_\theta^2 + \tilde{k}^2\right)\right\} \ .$$

Proof: Consider the function $\Phi = \left(\tilde{k}_\theta^2 + \tilde{k}^2\right)$. Suppose that $\Phi(\theta_0, \tau_1) = \sup(\tilde{k}_\theta^2 + k^2)$ and $\tau_1 > \tau_0$. We claim that $\tilde{k}_\theta$ must vanish at (θ_0, τ_1). For, if not, $\Phi_\theta = 2\tilde{k}_\theta \left(\tilde{k}_{\theta\theta} + \tilde{k}\right) = 0$. We must have $\tilde{k}_{\theta\theta} + \tilde{k} = 0$. Using

$\Phi_{\theta\theta} \leqslant 0$ and $\Phi_\tau \geqslant 0$ at (θ_0, τ_1), we have

$$0 \leqslant \tilde{k}\tilde{k}_\tau + \tilde{k}_\theta \tilde{k}_{\theta\tau}$$

$$\leqslant -\tilde{k}^2 + \tilde{k}_\theta \tilde{k}^2 \left(\tilde{k}_{\theta\theta\theta} + \tilde{k}_\theta\right) - \tilde{k}_\theta^2$$

$$\leqslant -\tilde{k}^2 - \tilde{k}_\theta^2 ,$$

which is impossible. So $\tilde{k}_\theta(\theta_0, \tau_1) = 0$. $\square$

Suppose that $\tilde{k}_{\max}(\tau) = \tilde{k}(\theta_0, \tau)$ and $\tilde{k}_{\max}(\tau) \geqslant \tilde{k}(\theta, \tau')$ for all $\tau' \in [\tau_0, \tau]$. It follows from this lemma that

$$\tilde{k}_{\max}(\tau) - \tilde{k}(\theta, \tau) \leqslant |\theta - \theta_0| \sup_\theta |\tilde{k}_\theta(\cdot, \tau)|$$

$$\leqslant |\theta - \theta_0| \left(C_0 + \tilde{k}_{\max}(\tau)\right) ,$$

where C_0 only depends on the initial curve. Therefore, for θ satisfying $|\theta - \theta_0| \leqslant 1/2$, we have

$$\tilde{k}_{\max}(\tau) \leqslant 2\tilde{k}(\theta, \tau) + C_0 . \tag{3.7}$$

Next, we derive an upper bound for the normalized curvature. The key step is the monotonicity of the entropy for the normalized flow. For any uniformly convex curve γ, its **entropy** is defined by

$$\mathcal{E}(\gamma) = \frac{1}{2\pi} \int_\gamma k \log k \, ds$$

$$= \fint \log k \, d\theta .$$

Lemma 3.8

$$\frac{d}{d\tau} \mathcal{E}\left(\tilde{\gamma}(\cdot, \tau)\right) \leqslant 0 .$$

Proof: By (3.6), we have

$$\frac{d\mathcal{E}}{d\tau} = \int \frac{\tilde{k}_\tau}{\tilde{k}} \ ,$$

and

$$\frac{d^2\mathcal{E}}{d\tau^2} = 2\left(\int \frac{\tilde{k}_\tau^2}{\tilde{k}^2} + \int \frac{\tilde{k}_\tau}{\tilde{k}} \right)$$

$$\geqslant 2\frac{d\mathcal{E}}{d\tau}\left(\frac{d\mathcal{E}}{d\tau} + 1 \right) .$$

Were $d\mathcal{E}/d\tau$ positive somewhere, it blows up at some later time. Since now $\mathcal{E}(\tau)$ is defined for all $\tau \geqslant \tau_0$, this is not possible. So $d\mathcal{E}/d\tau \leqslant 0$ everywhere. $\qquad\square$

The **width** of a convex curve in the direction $(\cos\theta, \sin\theta)$ is given by $w(\theta) = h(\theta) + h(\theta + \pi)$. We show that an upper bound on the entropy yields a positive lower bound for the width.

Lemma 3.9 *Let γ be a closed, convex curve and h its support function. There exists an absolute constant C_0 such that*

$$w(\theta) \geqslant C_0 e^{-\mathcal{E}(\gamma)}$$

for all θ.

Proof: For any fixed θ_0,

$$\int_{\theta_0}^{\pi+\theta_0} \frac{\sin(\theta - \theta_0)}{k}\, d\theta = \int_{\theta_0}^{\pi+\theta_0} \sin(\theta - \theta_0)(h_{\theta\theta} + h)\, d\theta$$

$$= w(\theta_0) ,$$

after performing integration by parts. By Jensen's inequality,

$$
\begin{aligned}
\log w(\theta_0) &= \log \frac{1}{\pi} \int_{\theta_0}^{\pi+\theta_0} \frac{|\sin(\theta - \theta_0)|}{k(\theta)} d\theta + \log \pi \\
&\geqslant \frac{1}{\pi} \int_0^\pi \log \sin \theta d\theta + \log \pi - \frac{1}{\pi} \int_0^\pi \log k(\theta + \theta_0) d\theta .
\end{aligned}
$$

A similar inequality can be obtained when we integrate from $\pi + \theta_0$ to $2\pi + \theta_0$. Hence, the lemma follows.   □

It follows from Lemmas 3.8 and 3.9 that the width of $\tilde\gamma(\cdot, t)$ in all directions are uniformly bounded from zero. Since the area enclosed by $\tilde\gamma$ is always equal to π, the length and diameter of $\tilde\gamma$ are bounded from above and its inradius has a positive lower bound. Now we proceed to bound the curvature of $\tilde\gamma$. If $\tilde k$ is unbounded, we can find $\{\tau_j\}$, $\tau_j \longrightarrow \infty$, such that $\tilde k_{\max}(\tau_j) \geqslant \tilde k_{\max}(\tau)$, for all $\tau \leqslant \tau_j$. Then for each $\tau = \tau_j$, we can use (3.7) to get

$$
\begin{aligned}
2\pi \mathcal{E}(0) &\geqslant \int \log \tilde k(\theta, \tau) d\theta \\
&\geqslant \int_{|\theta - \theta_0| \leqslant \frac{1}{2}} \log \tilde k(\theta, \tau) d\theta + \int_{\{\tilde k < 1\}} \tilde k \log \tilde k ds \\
&\geqslant \frac{1}{2} \log \frac{1}{2} + \frac{1}{2} \log \left(\tilde k_{\max}(\tau) - C_0 \right) - e^{-1} L ,
\end{aligned}
$$

where L is a bound on the length of $\tilde\gamma$. So the normalized curvature is bounded in $[\tau_0, \infty)$.

We still need to derive a positive lower bound for the curvature. In general, this may be done by using the Harnack's inequality (see next section). Here we use an elementary argument.

Define the functional

$$\mathcal{F}(\gamma) = \int \log h(\theta)d\theta ,$$

where we have assumed that the curve γ encloses the origin.

Lemma 3.10 *We have*

$$\frac{d}{d\tau}\mathcal{F}\big(\tilde{\gamma}(\cdot,\tau)\big) \leqslant 0 ,$$

and the equality holds if and only if $\tilde{\gamma}(\cdot,\tau)$ is the unit circle centered at the origin.

Proof: Use (3.5) and the Hölder inequality. □

Observe that $\log\tilde{h}$ could be very negative when the origin is very close to $\tilde{\gamma}$. Nevertheless, we claim that $\mathcal{F}\big(\tilde{\gamma}(\cdot,\tau)\big)$ is uniformly bounded from below. For, since $\tilde{k}$ is uniformly bounded, say, by M, and the origin is always inside $\tilde{\gamma}$, we can find a circle of radius $1/M$ which is contained inside $\tilde{\gamma}$ and passes through the origin. Without loss of generality, we may assume this circle is given by $\{(x,y) : (x+1/M)^2+y^2 = 1/M^2\}$. Then $\tilde{h}(\cdot,\tau)$ is greater than $M^{-1}(1-\cos\theta)$, the support function of this circle. As a result,

$$\mathcal{F}\big(\tilde{\gamma}(\cdot,\tau)\big) \;\geqslant\; -\log M + \int \log(1-\cos\theta)d\theta$$

$$> \;-\infty .$$

Now, consider the functional

$$\mathcal{I}\big(\tilde{\gamma}(\cdot,t)\big) = \int \frac{1}{2}\big(\tilde{h}^2 - \tilde{h}_\theta^2\big)d\theta - \int \log\tilde{h}\,d\theta .$$

From the above discussion, we know that it is bounded from below. By (3.5),

$$\mathcal{I}\big(\tilde{\gamma}(\cdot,\tau_0)\big) - \mathcal{I}\big(\tilde{\gamma}(\cdot,\tau)\big) = \int_{\tau_0}^{\tau}\!\!\int \frac{(\tilde{h}-\tilde{k})^2}{\tilde{h}\tilde{k}}(\theta,\tau)d\theta d\tau \qquad (3.8)$$

for all $\tau \in [\tau_0, \infty)$. For each $[j, j+1]$, we can find $s_j \in [j, j+1]$ such that

$$\lim_{j \longrightarrow \infty} \int \frac{(\tilde{h} - \tilde{k})^2}{\tilde{h}\tilde{k}}(\theta, s_j)d\theta = 0 \;.$$

Let $\{\tilde{\gamma}(\cdot, s_{j_k})\}$ be a subsequence of $\{\tilde{\gamma}(\cdot, s_j)\}$ which converges to some convex curve $\tilde{\gamma}$. The support function of $\tilde{\gamma}$, $\tilde{h}$ is non-negative and $\{\tilde{h}(\cdot, s_{j_k})\}$ converges uniformly to $\tilde{h}$ as $s_{j_k} \longrightarrow \infty$. Let I be an interval on which $\tilde{h}$ is positive and let φ be any test function. We have

$$\left| \int_I \left[\tilde{h}(\cdot, s_{j_k})\varphi_{\theta\theta} + \left(\tilde{h}(\cdot, s_{j_k}) - \frac{1}{\tilde{h}(\cdot, s_{j_k})} \right)\varphi \right] d\theta \right|$$

$$= \left| \int_I \frac{1}{\tilde{h}(\cdot, \tau_{j_k})} \left(\frac{\tilde{h}(\cdot, s_{j_k})}{\tilde{k}(\cdot, s_{j_k})} - 1 \right)\varphi d\theta \right|$$

$$\leqslant \sup_I \frac{|\varphi|}{\tilde{h}(\cdot, s_{j_k})} \left(\int \frac{\tilde{h}(\cdot, s_{j_k})}{\tilde{k}(\cdot, s_{j_k})} d\theta \right)^{1/2} \left(\int \frac{\tilde{k}(\cdot, s_{j_k})}{\tilde{h}(\cdot, s_{j_k})} \left(\frac{\tilde{h}(\cdot, s_{j_k})}{\tilde{k}(\cdot, s_{j_k})} - 1 \right)^2 d\theta \right)^{1/2}$$

$$\longrightarrow 0 \;, \quad \text{as} \quad s_{j_k} \longrightarrow \infty \;.$$

Hence, $\tilde{h}$ is a weak solution of $h_{\theta\theta} + h = 1/h$. By standard regularity theory, $\tilde{h}$ is smooth. Furthermore, by our discussion in §2.4, $\tilde{h}$ must be positive everywhere and Proposition 2.4 asserts that $\tilde{h} \equiv 1$. In conclusion, we have shown that every convergent subsequence of $\{\tilde{\gamma}(\cdot, s_j)\}$ must tend to the unit circle, and so $\tilde{\gamma}(\cdot, s_j)$ converges to the unit circle. As the curvature and its gradient are bounded, $\{\tilde{k}(\cdot, s_j)\}$ converges uniformly to 1.

Now we show the curvature has a positive lower bound. For any τ, we can find j such that $\tau \in [s_j, s_{j+1}]$. From the differential inequality

$$\frac{d\tilde{k}_{\min}}{d\tau} \geqslant \tilde{k}_{\min}(\tilde{k}_{\min}^2 - 1) \;, \quad \tilde{k}_{\min} < 1 \;,$$

we have

$$\log \frac{\sqrt{1 - \widetilde{k}^2_{\min}(\tau)}}{\widetilde{k}_{\min}(\tau)} \leqslant \log \frac{\sqrt{1 - \widetilde{k}^2_{\min}(\tau_j)}}{\widetilde{k}_{\min}(\tau_j)} + \tau_{j+1} - \tau_j \ .$$

Using $s_{j+1} - s_j \leqslant 2$ and $\widetilde{k}_{\min}(\tau_j)$ tends to 1 as $j \longrightarrow \infty$, we conclude that $\widetilde{k}_{\min}(\tau)$ has a positive lower bound.

With two-sided bounds on the curvature, we can use standard parabolic regularity to obtain all higher order bounds. It follows from (3.8) that

$$\lim_{\tau \longrightarrow 0} \int \frac{(\widetilde{h} - \widetilde{k})^2}{\widetilde{h}\widetilde{k}}(\theta, \tau) d\theta = 0$$

So $\widetilde{h}(\cdot, \tau)$ converges to 1 smoothly.

To finish the proof, we show the convergence is exponential.

Lemma 3.11 *For any $\varepsilon > 0$, there exists $\tau_1 \geqslant \tau_0$ such that*

$$\int \widetilde{k}^2_{\theta\theta} \geqslant (4 - \varepsilon) \int \widetilde{k}^2_\theta \ , \ for \ \tau \geqslant \tau_1 \ .$$

Proof: As $(1/\widetilde{k})_\theta$ is orthogonal to 1, $\sin\theta$ and $\cos\theta$,

$$\int \left(\frac{1}{\widetilde{k}}\right)^2_{\theta\theta} \geqslant 4 \int \left(\frac{1}{\widetilde{k}}\right)^2_\theta \ .$$

The desired inequality follows from the fact that $\widetilde{k}$ tends to 1 smoothly.
□

By Lemma 3.11, for large τ, we have

$$\frac{1}{2}\frac{d}{d\tau}\int \widetilde{k}^2_\theta \ = \ -\int \widetilde{k}^2\widetilde{k}^2_{\theta\theta} + \int (3\widetilde{k}^2 - 1)\widetilde{k}^2_\theta$$

$$\leqslant \ -(2 - \varepsilon)\int \widetilde{k}^2_\theta \ .$$

Therefore,

$$\int \widetilde{k}^2_\theta \leqslant e^{2(2-\varepsilon)(\tau_1 - \tau)} \int \widetilde{k}^2_\theta(\cdot, \tau_1) \ .$$

For each τ, we can find θ_0 such that $\tilde{k}(\theta_0, \tau) = 1$. So, for any $\varepsilon > 0$, there exists C_ε and τ_ε such that

$$\left| \tilde{k}(\theta, \tau) - 1 \right| \leqslant C_\varepsilon \, e^{-(2-\varepsilon)\tau} \quad , \quad \text{all} \quad \tau \geqslant \tau_\varepsilon .$$

3.4   The contracting case of the ACEF

In this section, we begin the study of the ACEF, which is now placed in the following form,

$$\frac{\partial \gamma}{\partial t} = \left(\Phi(\theta)k + \lambda \Psi(\theta) \right) n , \qquad (3.9)_\lambda$$

where Φ and Ψ are positive 2π-periodic functions of the normal angle and $\lambda \in \mathbb{R}$. When $\Phi \equiv 1$, $\Psi \equiv 1$, and γ_0 is a circle of radius R, the flow (3.9) shrinks to a point (resp. expands to infinity) when $1/R + \lambda > 0$ (resp. $1/R + \lambda < 0$), and it is stationary when $1/R + \lambda = 0$. It turns out this behaviour is typical.

In fact, by Propositions 3.2–3.4, we know that any solution of $(3.9)_\lambda$ starting at a closed, convex curve is uniformly convex for each t, and ω is finite if and only if it shrinks to a point. Using the strong separation principle, we also know the following facts hold:

(a) Let γ_i be solution of $(3.9)_{\lambda_i}$, $i = 1, 2$, satisfying $\gamma_1(\cdot, 0) = \gamma_2(\cdot, 0)$, and $\lambda_1 < \lambda_2$. Then, $\gamma_2(\cdot, t)$ is enclosed by $\gamma_1(\cdot, t)$ as long as $\gamma_2(\cdot, t)$ exists.

(b) If $\gamma(\cdot, t)$ is enclosed in a circle of radius R, $1/R \geqslant -\lambda \Psi_{\max}/\Phi_{\min}$, at some t, it shrinks to a point; if $\gamma(\cdot, t)$ encloses a circle of radius R, $> -\Phi_{\max}/(\lambda \Psi_{\min})$, it expands to infinity.

Let γ_0 be a fixed closed, convex curve and $\gamma(\cdot, t)$ the solution of $(3.9)_\lambda$ starting at γ_0. Let

$$A = \left\{ \lambda : \gamma(\cdot, t) \text{ shrinks to a point} \right\} ,$$

and

$$B = \left\{ \lambda : \gamma(\cdot, t) \text{ expands to infinity} \right\}.$$

From the above facts, we know that A and B are open intervals of the form (λ^*, ∞) and $(-\infty, \lambda_*)$ where $\lambda^* \geqslant \lambda_*$ and $\lambda^* < 0$. Now we can state our main result.

Theorem 3.12 *Consider* $(3.9)_\lambda$, *where* Φ *and* Ψ *are positive and* $\lambda \in \mathbb{R}$. *For any closed, convex* γ_0, *there exists* $\lambda^* < 0$ *such that the Cauchy problem for* $(3.9)_\lambda$ *has a unique solution* $\gamma(\cdot, t)$ *which is uniformly convex for each* t *in* $(0, \omega)$. *Moreover,* $\lambda^* = \lambda_*$ *and the following statements hold:*

(i) *When* $\lambda > \lambda^*$, ω *is finite and* $\gamma(\cdot, t)$ *contracts to a point as* $t \uparrow \omega$. *Moreover, if we normalize* $\gamma(\cdot, t)$ *so that its enclosed area is constant, the normalized flow subconverges to self-similar solutions of the flow*

$$\frac{\partial \gamma}{\partial t} = \Phi(\theta) k \boldsymbol{n} . \tag{3.10}$$

(ii) *When* $\lambda = \lambda^*$, $\omega = \infty$ *and the curvature of* $\gamma(\cdot, t)$ *is uniformly bounded. The flow converges smoothly to a stationary solution of* $(3.9)_\lambda$ *if and only if there exists a* 2π-*periodic function* ξ *such that*

$$\frac{\Phi}{\Psi} = \frac{d^2 \xi}{d\theta^2} + \xi .$$

(iii) *When* $\lambda < \lambda^*$, $\omega = \infty$ *and the flow expands to infinity as* $t \longrightarrow \infty$. *If the polar diagram of* $1/\Psi$ *is uniformly convex, i.e.,* $\Psi + d^2\Psi/d\theta^2 > 0$ *for all* θ, *then* $\gamma(\cdot, t)/t$ *converges smoothly to the boundary of the Wulff region of* $-\lambda\Psi$.

The **Wulff region** of Ψ is given by $\{(x, y), \langle(x, y), (\cos\theta, \sin\theta)\rangle \leqslant \Psi(\theta)\}$.

In the rest of this section we prove Theorem 3.12 (i). Let $F = \Phi k + \lambda \Psi$. Then F satisfies

$$\Phi^{-1} \frac{\partial F}{\partial t} = k^2 (F_{\theta\theta} + F) . \qquad (3.11)$$

We begin with a gradient estimate for F.

Lemma 3.13 *Either*

$$\frac{\partial F}{\partial t}(\theta, t) \geqslant 0$$

or

$$\left(F^2 + F_\theta^2\right)(\theta, t) \leqslant M^2 ,$$

where $M^2 = \max \left\{ \sup \left(F^2 + F_\theta^2\right)(\theta, 0), \sup \Psi \right\}$.

Proof: For any fixed (θ_0, t_0), $t_0 > 0$, let $B = \left(F^2 + F_\theta^2\right)^{1/2}(\theta_0, t_0)$. Suppose that $B > M$. We are going to show that $(F_{\theta\theta} + F) \geqslant 0$ at (θ_0, t_0).

Choose $\xi \in (-\pi, \pi)$ such that

$$F(\theta_0, t_0) = B \cos \xi , \quad \frac{\partial F}{\partial \theta}(\theta_0, t_0) = B \sin \xi .$$

Consider the function $G = F - F^*$, where $F^* = B \cos(\theta - \theta_0 + \xi)$ is a stationary solution of (3.11). By assumption, G is positive at $(\theta_0 - \xi \pm \pi, t)$, $0 \leqslant t \leqslant t_0$, and has a double zero at (θ_0, t_0). Since $B > M$, $G(\theta, 0)$ must vanish somewhere in $(\theta_0 - \xi - \pi, \theta_0 - \xi + \pi)$.

Suppose that θ_1 is a root in $(\theta_0 - \xi - \pi, \theta_0 - \xi)$. We have

$$
\begin{aligned}
\frac{\partial G}{\partial \theta}(\theta_1, 0) &= \frac{\partial F}{\partial \theta}(\theta_1, 0) + B \sin(\theta_1 - \theta_0 + \xi) \\[2mm]
&\leqslant \left(M^2 - F^2(\theta_1, 0) \right)^{1/2} - B \left| \sin(\theta_1 - \theta_0 + \xi) \right| \\[2mm]
&< \left(B^2 - F^{*2}(\theta_1) \right)^{1/2} - B \left| \sin(\theta_1 - \theta_0 + \xi) \right| \\[2mm]
&= 0 .
\end{aligned}
$$

Similarly, we can show that $\partial G/\partial \theta(\theta_2, 0) > 0$ for any root θ_2 in $(\theta_0 - \xi, \theta_0 - \xi + \pi)$. Therefore, $G(\cdot, 0)$ has exactly two roots in $(\theta_0 - \xi - \pi, \theta_0 - \xi + \pi)$. By the Sturm oscillation theorem, $G(\cdot, t_0)$ has no roots other than θ_0. So $F_{\theta\theta} + F = G_{\theta\theta} \geqslant 0$ at (θ_0, t_0). $\qquad\square$

We can deduce a gradient estimate from this lemma. In fact, let's assume $(F_{\theta\theta} + F)(\theta_1, t_1) > 0$ and $F_\theta(\theta_1, t_1) > 0$. We can find an interval (θ_1, θ_2) on which $F_{\theta\theta} + F$ is nonnegative and $\left| F_\theta(\theta_2, t_1) \right| \leqslant M$. Therefore, we have

$$
F_\theta(\theta_1, t_1) \leqslant F_\theta(\theta_2, t_1) + \int_{\theta_1}^{\theta_2} F \, d\theta ,
$$

which implies

$$
\sup_\theta \left| F_\theta(\theta, t) \right| \leqslant M + \int_0^{2\pi} \left| F(\theta, t) \right| d\theta . \tag{3.12}
$$

It follows immediately that

$$
\begin{aligned}
F_{\max}(t) &\leqslant \fint F(\theta, t) d\theta + 2\pi \sup \left| F_\theta(\cdot, t) \right| \\[2mm]
&\leqslant M_1 \left(1 + \int_0^{2\pi} \left| F(\theta, t) \right| d\theta \right) , \tag{3.13}
\end{aligned}
$$

and

$$F_{\max}(t) \leqslant 2F(\theta, t) + \frac{M}{2\pi} \qquad (3.14)$$

for all θ, $|\theta - \theta_0| \leqslant 1/(4\pi)$ where $F_{\max}(t) = F(\theta_0, t)$.

Now we study the normalized flow. First, from (1.19), we have

$$\lim_{t \uparrow \omega} \frac{A(t)}{2(\omega - t)} = \frac{1}{2} \int_0^{2\pi} \Phi(\theta) d\theta .$$

So we let

$$\tilde{\gamma}(\cdot, t) = (2\omega - 2t)^{-1/2} \gamma(\cdot, t) .$$

The area enclosed by $\tilde{\gamma}$ approaches the constant $\frac{1}{2} \int_0^{2\pi} \Phi d\theta$ as $t \uparrow \omega$. Changing the time scale from t to $2\tau = -\log(1 - \omega^{-1}t)$, the equations for the normalized support function and the normalized curvature are given, respectively, by

$$\frac{\partial \tilde{h}}{\partial \tau} = -\left(\Phi \tilde{k} + \sqrt{2\omega}\, e^{-\tau} \lambda \Psi\right) + \tilde{h} , \qquad (3.15)$$

$$\frac{\partial \tilde{k}}{\partial \tau} = \tilde{k}^2 \left(\frac{\partial^2 (\Phi k)}{\partial \theta^2} + \Phi \tilde{k}\right) - \tilde{k} + \sqrt{2\omega}\, e^{-\tau} \lambda \tilde{k}^2 \left(\Psi_{\theta\theta} + \Psi\right) . \quad (3.16)$$

The entropy for the normalized flow is defined by

$$\mathcal{E}\left(\tilde{\gamma}(\cdot, \tau)\right) = \int \Phi(\theta) \log \tilde{k}(\theta, \tau) d\theta . \qquad (3.17)$$

We shall show that it is uniformly bounded for all $\tau \in [0, \infty)$.

We have

$$\frac{d}{d\tau}\widetilde{\mathcal{E}}(\tau) \tag{3.18}$$

$$= \int_0^{2\pi} \frac{\Phi(\theta)}{\widetilde{k}(\theta,\tau)} \frac{\partial \widetilde{k}}{\partial \tau}(\theta,\tau)d\theta$$

$$= \int_0^{2\pi} \left\{ \Phi\widetilde{k}\left(\frac{\partial^2(\Phi\widetilde{k})}{\partial\theta^2} + \Phi\widetilde{k}\right) - \Phi + \sqrt{2\omega}e^{-\tau}\Phi\widetilde{k}\lambda\left(\Psi_{\theta\theta} + \Psi\right) \right\} d\theta$$

$$= \int_0^{2\pi} u(\theta,\tau)d\theta + \int_0^{2\pi} \left[\sqrt{2\omega}e^{-\tau}\Phi\widetilde{k}\lambda\left(\Psi_{\theta\theta} + \Psi\right) + 2\Lambda e^{-\tau}\widetilde{k}\Phi \right] d\theta \; ,$$

where

$$u = \Phi\widetilde{k}\left(\frac{\partial^2(\Phi\widetilde{k})}{\partial\theta^2} + \Phi\widetilde{k}\right) - \Phi - 2\Lambda e^{-\tau}\Phi\widetilde{k} \; ,$$

and $\Lambda > 0$ to be specified later. Let's compute

$$\frac{d}{d\tau}\int_0^{2\pi} u\,d\theta$$

$$= \frac{d}{d\tau}\int_0^{2\pi} \left[\left(\Phi\widetilde{k}\right)^2 - \left(\frac{\partial(\Phi\widetilde{k})}{\partial\theta}\right)^2 - \Phi - 2\Lambda e^{-\tau}\widetilde{k}\Phi \right] d\theta$$

$$= \int_0^{2\pi} \left\{ 2\left(\Phi\widetilde{k}\right)\left(\Phi\widetilde{k}_\tau\right) - 2\left(\frac{\partial(\Phi\widetilde{k})}{\partial\theta}\right)\left(\frac{\partial(\Phi\widetilde{k}_\tau)}{\partial\theta}\right) + 2\Lambda e^{-\tau}\widetilde{k}\Phi \right.$$

$$\left. - 2\Lambda e^{-\tau}\widetilde{k}_\tau\Phi \right\} d\theta$$

$$= 2 \int_0^{2\pi} \left[\left(\frac{\partial^2 (\Phi \tilde{k})}{\partial \theta^2} + \Phi \tilde{k} \right) \Phi \tilde{k}_\tau + \Lambda e^{-\tau} \tilde{k} \Phi - \Lambda e^{-\tau} \tilde{k}_\tau \Phi \right] d\theta$$

$$= 2 \int_0^{2\pi} \left(\frac{\partial^2 (\Phi \tilde{k})}{\partial \theta^2} + \Phi \tilde{k} \right) \Phi \left[\tilde{k}^2 \left(\frac{\partial^2 (\Phi \tilde{k})}{\partial \theta^2} + \Phi \tilde{k} \right) - \tilde{k} \right.$$

$$\left. + \sqrt{2\omega} e^{-\tau} \tilde{k}^2 \lambda \left(\Psi_{\theta\theta} + \Psi \right) \right] d\theta + \int_0^{2\pi} 2\Lambda e^{-\tau} \tilde{k} \Phi d\theta$$

$$- 2\Lambda \int_0^{2\pi} e^{-\tau} \Phi \left[\tilde{k}^2 \left(\frac{\partial^2 (\Phi \tilde{k})}{\partial \theta^2} + \Phi \tilde{k} \right) - \tilde{k} \right.$$

$$\left. + \sqrt{2\omega} e^{-\tau} \tilde{k}^2 \lambda \left(\Psi_{\theta\theta} + \Psi \right) \right] d\theta$$

$$= 2 \int_0^{2\pi} \left\{ \frac{1}{\Phi} \left[\Phi \tilde{k} \left(\frac{\partial^2 (\Phi \tilde{k})}{\partial \theta^2} + \Phi \tilde{k} \right) \right]^2 - \Phi \tilde{k} \left(\frac{\partial^2 (\Phi \tilde{k})}{\partial \theta^2} + \Phi \tilde{k} \right) \right.$$

$$\left. + \sqrt{2\omega} e^{-\tau} \Phi \tilde{k}^2 \left(\frac{\partial^2 (\Phi \tilde{k})}{\partial \theta^2} + \Phi \tilde{k} \right) \lambda \left(\Psi_{\theta\theta} + \Psi \right) \right\} d\theta$$

$$+ 2\Lambda \int_0^{2\pi} e^{-\tau} \tilde{k} \Phi d\theta - 2\Lambda \int_0^{2\pi} e^{-\tau} \Phi \tilde{k}^2 \left(\frac{\partial^2 (\Phi \tilde{k})}{\partial \theta^2} + \Phi \tilde{k} \right) d\theta$$

$$+ 2\Lambda \int_0^{2\pi} e^{-\tau} \tilde{k} \Phi d\theta - 2\Lambda \int_0^{2\pi} \sqrt{2\omega} e^{-2\tau} \Phi \tilde{k}^2 \lambda \left(\Psi_{\theta\theta} + \Psi \right) d\theta$$

$$= 2 \int_0^{2\pi} \frac{1}{\Phi} \left\{ (\Phi \tilde{k})^2 \left(\frac{\partial^2 (\Phi \tilde{k})}{\partial \theta^2} + \Phi \tilde{k} \right)^2 - 2\Phi \tilde{k} \left(\frac{\partial^2 (\Phi \tilde{k})}{\partial \theta^2} + \Phi \tilde{k} \right) \Phi \right.$$

$$+ \Phi \tilde{k} \left(\frac{\partial^2 (\Phi \tilde{k})}{\partial \theta^2} + \Phi \tilde{k} \right) \Phi - 4\Lambda e^{-\tau} (\Phi \tilde{k})^2 \left(\frac{\partial^2 (\Phi \tilde{k})}{\partial \theta^2} + \Phi \tilde{k} \right)$$

$$+ 4\Lambda e^{-\tau} (\Phi \tilde{k})^2 \left(\frac{\partial^2 (\Phi \tilde{k})}{\partial \theta^2} + \Phi \tilde{k} \right) + 4\Lambda e^{-\tau} \tilde{k} \Phi^2 - 4\Lambda e^{-\tau} \tilde{k} \Phi^2$$

$$+ \Phi^2 - \Phi^2 + 4\Lambda^2 e^{-2\tau} (\tilde{k} \Phi)^2 - 4\Lambda^2 e^{-2\tau} (\tilde{k} \Phi)^2 \right\} d\theta$$

$$+ 2 \int_0^{2\pi} \sqrt{2\omega} e^{-\tau} \Phi \tilde{k}^2 \left(\frac{\partial^2 (\Phi \tilde{k})}{\partial \theta^2} + \Phi \tilde{k} \right) \lambda \left(\Psi_{\theta\theta} + \Psi \right) d\theta$$

$$+2\Lambda \int_0^{2\pi} e^{-\tau}\Phi\widetilde{k}d\theta - 2\Lambda \int_0^{2\pi} e^{-\tau}\Phi\widetilde{k}^2 \left(\frac{\partial^2(\Phi\widetilde{k})}{\partial\theta^2} + \Phi\widetilde{k} \right) d\theta$$

$$+2\Lambda \int_0^{2\pi} e^{-\tau}\Phi\widetilde{k}d\theta - 2\Lambda\sqrt{2\omega} \int_0^{2\pi} e^{-2\tau}\Phi\widetilde{k}^2 \lambda \left(\Psi_{\theta\theta} + \Psi \right) d\theta$$

$$= \; 2\int_0^{2\pi} \frac{1}{\Phi}u^2 d\theta + 2\int_0^{2\pi} \Phi\widetilde{k} \left(\frac{\partial^2(\Phi\widetilde{k})}{\partial\theta^2} + \Phi\widetilde{k} \right) d\theta +$$

$$8\Lambda \int_0^{2\pi} e^{-\tau}\Phi\widetilde{k}^2 \left(\frac{\partial^2(\Phi\widetilde{k})}{\partial\theta^2} + \Phi\widetilde{k} \right) d\theta - 8\Lambda \int_0^{2\pi} e^{-\tau}\Phi\widetilde{k}d\theta$$

$$-2\int_0^{2\pi} \Phi d\theta - 8\Lambda^2 \int_0^{2\pi} e^{-2\tau}\Phi\widetilde{k}^2 d\theta +$$

$$2\int_0^{2\pi} \sqrt{2\omega}e^{-\tau}\Phi\widetilde{k}^2 \left(\frac{\partial^2(\Phi\widetilde{k})}{\partial\theta^2} + \Phi\widetilde{k} \right) \lambda \left(\Psi_{\theta\theta} + \Psi \right) d\theta +$$

$$+4\Lambda \int_0^{2\pi} e^{-\tau}\Phi\widetilde{k}d\theta - 2\Lambda \int_0^{2\pi} e^{-\tau}\Phi\widetilde{k}^2 \left(\frac{\partial^2(\Phi\widetilde{k})}{\partial\theta^2} + \Phi\widetilde{k} \right) d\theta$$

$$-2\Lambda\sqrt{2\omega} \int_0^{2\pi} e^{-2\tau}\Phi\widetilde{k}^2 \lambda \left(\Psi_{\theta\theta} + \Psi \right) d\theta$$

$$= \; 2\int_0^{2\pi} \frac{1}{\Phi}u^2 d\theta + 2\int_0^{2\pi} u d\theta + 6\Lambda \int_0^{2\pi} e^{-\tau}\Phi\widetilde{k}^2 \left(\frac{\partial^2(\Phi\widetilde{k})}{\partial\theta^2} + \Phi\widetilde{k} \right) d\theta$$

$$-8\Lambda^2 \int_0^{2\pi} e^{-2\tau}\Phi\widetilde{k}^2 d\theta$$

$$+2\sqrt{2\omega} \int_0^{2\pi} e^{-\tau}\Phi\widetilde{k}^2 \left(\frac{\partial^2(\Phi\widetilde{k})}{\partial\theta^2} + \Phi\widetilde{k} \right) \lambda \left(\Psi_{\theta\theta} + \Psi \right) d\theta$$

$$-2\Lambda\sqrt{2\omega} \int_0^{2\pi} e^{-2\tau}\Phi\widetilde{k}^2 \lambda \left(\Psi_{\theta\theta} + \Psi \right) d\theta$$

$$= \; 2\int_0^{2\pi} \frac{1}{\Phi}u^2 d\theta + 2\int_0^{2\pi} u d\theta$$

$$+4\Lambda \int_0^{2\pi} e^{-\tau}\widetilde{k} \left[\Phi\widetilde{k} \left(\frac{\partial^2(\Phi\widetilde{k})}{\partial\theta^2} + \Phi\widetilde{k} \right) - \Phi - 2\Lambda e^{-\tau}\Phi\widetilde{k} \right] d\theta$$

$$+4\Lambda \int_0^{2\pi} e^{-\tau}\Phi\widetilde{k}d\theta + 8\Lambda^2 \int_0^{2\pi} e^{-2\tau}\Phi\widetilde{k}^2 d\theta +$$

$$2\int_0^{2\pi} e^{-\tau}\Phi\widetilde{k}^2 \left(\frac{\partial^2(\Phi\widetilde{k})}{\partial\theta^2} + \Phi\widetilde{k} \right) \left[\Lambda + \sqrt{2\omega}\lambda \left(\Psi_{\theta\theta} + \Psi \right) \right] d\theta$$

$$-8\Lambda^2 \int_0^{2\pi} e^{-2\tau} \Phi \widetilde{k}^2 \, d\theta - 2\Lambda\sqrt{2\omega} \int_0^{2\pi} e^{-2\tau} \Phi \widetilde{k}^2 \lambda \left(\Psi_{\theta\theta} + \Psi\right) d\theta$$

$$= 2 \int_0^{2\pi} \frac{1}{\Phi} u^2 \, d\theta + 2 \int_0^{2\pi} (1 + 2\Lambda e^{-\tau}\widetilde{k}) u \, d\theta + 4\Lambda \int_0^{2\pi} e^{-\tau}\Phi\widetilde{k} d\theta$$

$$+ 2 \int_0^{2\pi} e^{-\tau} \Phi \widetilde{k}^2 \left(\frac{\partial^2 (\Phi\widetilde{k})}{\partial\theta^2} + \Phi\widetilde{k} \right) \left[\Lambda + \sqrt{2\omega}\lambda \left(\Psi_{\theta\theta} + \Psi\right)\right] d\theta$$

$$-2\Lambda\sqrt{2\omega} \int_0^{2\pi} e^{-2\tau} \Phi \widetilde{k}^2 \lambda \left(\Psi_{\theta\theta} + \Psi\right) d\theta.$$

Using

$$2 \int_0^{2\pi} 2\Lambda e^{-\tau}\widetilde{k} u \, d\theta \;=\; 2 \int_0^{2\pi} 2\Lambda e^{-\tau}\Phi^{1/2}\widetilde{k}\Phi^{-1/2} u \, d\theta$$

$$\geqslant \; -\int_0^{2\pi} \frac{1}{\Phi} u^2 \, d\theta - 4 \int_0^{2\pi} \Lambda^2 e^{-2\tau}\Phi\widetilde{k}^2 \, d\theta,$$

we get

$$\frac{d}{d\tau} \int_0^{2\pi} u \, d\theta \;\geqslant\; \int_0^{2\pi} \frac{1}{\Phi} u^2 \, d\theta + \int_0^{2\pi} 2u \, d\theta - 4\Lambda^2 \int_0^{2\pi} e^{-2\tau}\Phi\widetilde{k}^2 \, d\theta$$

$$+ 4\Lambda \int_0^{2\pi} e^{-\tau}\Phi\widetilde{k} d\theta + 2 \int_0^{2\pi} e^{-\tau}\Phi\widetilde{k}^2$$

$$\left(\frac{\partial^2 (\Phi\widetilde{k})}{\partial\theta^2} + \Phi\widetilde{k} \right) \left[\Lambda + + \sqrt{2\omega}\lambda \left(\Psi_{\theta\theta} + \Psi\right)\right] d\theta$$

$$- 2\Lambda\sqrt{2\omega} \int_0^{2\pi} e^{-2\tau}\Phi\widetilde{k}^2 \lambda \left(\Psi_{\theta\theta} + \Psi\right) d\theta.$$

Now, choosing

$$\Lambda = \sqrt{2\omega}\, \max_\theta |\lambda|\, |\Psi_{\theta\theta} + \Psi|\ ,$$

we have

$$\frac{d}{d\tau} \int_0^{2\pi} u \, d\theta$$

$$\geqslant \int_0^{2\pi} \frac{1}{\Phi} u^2 \, d\theta + \int_0^{2\pi} 2u \, d\theta - 6\Lambda^2 \int_0^{2\pi} e^{-2\tau}\Phi\widetilde{k}^2 \, d\theta$$

$$+ \; 4\Lambda \int_0^{2\pi} e^{-\tau}\Phi\widetilde{k} d\theta$$

$$+ \; 2 \int_0^{2\pi} e^{-\tau}\Phi\widetilde{k}^2 \left(\frac{\partial^2 (\Phi\widetilde{k})}{\partial\theta^2} + \Phi\widetilde{k} \right) \left[\Lambda + \sqrt{2\omega}\lambda \left(\Psi_{\theta\theta} + \Psi\right)\right] d\theta.$$

The last term in the right hand side of this inequality can be estimated as follows:

$$2 \int_0^{2\pi} e^{-\tau} \Phi \widetilde{k}^2 \left(\frac{\partial^2 (\Phi \widetilde{k})}{\partial \theta^2} + \Phi \widetilde{k} \right) \left[\Lambda + \sqrt{2\omega}\lambda \left(\Psi_{\theta\theta} + \Psi \right) \right] d\theta$$

$$= 2 \int_0^{2\pi} e^{-\tau} \widetilde{k} u \left[\Lambda + \sqrt{2\omega}\lambda \left(\Psi_{\theta\theta} + \Psi \right) \right] d\theta +$$

$$2 \int_0^{2\pi} e^{-\tau} \widetilde{k} (\Phi + 2\Lambda e^{-\tau} \Phi \widetilde{k}) \left[\Lambda + \sqrt{2\omega}\lambda \left(\Psi_{\theta\theta} + \Psi \right) \right] d\theta$$

$$\geqslant 2 \int_0^{2\pi} \frac{u}{\sqrt{2\Phi}} \sqrt{2\Phi} e^{-\tau} \widetilde{k} \left[\Lambda + \sqrt{2\omega}\lambda \left(\Psi_{\theta\theta} + \Psi \right) \right] d\theta$$

$$\geqslant -\frac{1}{2} \int_0^{2\pi} \frac{1}{\Phi} u^2 d\theta - 2 \int_0^{2\pi} e^{-2\tau} \Phi \widetilde{k}^2 \left[\Lambda + \sqrt{2\omega}\lambda \left(\Psi_{\theta\theta} + \Psi \right) \right]^2 d\theta$$

$$\geqslant -\frac{1}{2} \int_0^{2\pi} \frac{1}{\Phi} u^2 d\theta - 8\Lambda^2 \int_0^{2\pi} \Phi e^{-2\tau} \widetilde{k}^2 d\theta.$$

Finally, we arrive at

$$\frac{d}{d\tau} \int_0^{2\pi} u d\theta \geqslant \frac{1}{2} \int_0^{2\pi} \frac{1}{\Phi} u^2 d\theta + 2 \int_0^{2\pi} u d\theta$$

$$- 14\Lambda^2 \int_0^{2\pi} e^{-2\tau} \Phi \widetilde{k}^2 d\theta + 4\Lambda \int_0^{2\pi} e^{-\tau} \Phi \widetilde{k} d\theta . \quad (3.19)$$

Lemma 3.14

$$\sqrt{2\omega} \int_0^\infty \int_0^{2\pi} \Phi(\theta) e^{-\tau} \widetilde{k}(\theta, \tau) d\theta d\tau \leqslant L(0) + \omega \int_0^{2\pi} |\lambda \Psi(\theta)| d\theta.$$

Proof: Integrate the equation

$$\frac{dL}{d\tau} = \frac{dL}{dt} \frac{dt}{d\tau}$$

$$= -\sqrt{2\omega} \int_0^{2\pi} \Phi(\theta) e^{-\tau} \widetilde{k}(\theta, \tau) d\theta - 2\omega e^{-2\tau} \int_0^{2\pi} \lambda \Psi(\theta) d\theta.$$

□

To dispose the last term on the right hand side of (3.19), we need the following lemma:

Lemma 3.15 *We have*

$$e^{-\tau} \widetilde{k}_{\max}(\tau) \longrightarrow 0 \quad as \quad \tau \longrightarrow \infty .$$

Proof: From

$$\frac{dL}{d\tau} = -\sqrt{2\omega} \int_0^{2\pi} \Phi(\theta) e^{-\tau} \widetilde{k}(\theta, \tau) d\theta$$

$$-2\omega e^{-2\tau} \int_0^{2\pi} \lambda \Psi(\theta) d\theta \ ,$$

we have

$$\int_0^\infty \int_0^{2\pi} e^{-\tau} \widetilde{k}(\theta, \tau) d\theta d\tau \leqslant C_1 \ , \tag{3.20}$$

for some constant C_1. It follows from (3.13) and (3.20) that

$$\int_0^\infty F_{\max}(\tau) e^{-2\tau} d\tau \leqslant C_2$$

for some constant C_2. As $F_{\max}(\tau)$ tends to infinity as $\tau \longrightarrow \infty$, we can use Lemma 3.13 to show that $F_{\max}$ is non-decreasing for large τ. As a result,

$$F_{\max}(\tau) e^{-2\tau} \leqslant 2 \int_\tau^\infty F_{\max}(s) e^{-2s} ds$$

$$\longrightarrow 0 \ , \quad \text{as} \quad \tau \longrightarrow \infty \ .$$

$\square$

By this lemma, we finally obtain

$$\frac{d}{d\tau} \int_0^{2\pi} u d\theta \ \geqslant \ \frac{1}{2} \int_0^{2\pi} \frac{u^2}{\Phi} d\theta + 2 \int_0^{2\pi} u d\theta$$

$$\geqslant \ \frac{1}{2 \int \Phi d\theta} \left(\int_0^{2\pi} u d\theta \right)^2 + 2 \int_0^{2\pi} u d\theta$$

for all large τ. As before, it means that $\displaystyle\int_0^{2\pi} u\,d\theta \leqslant 0$ eventually, and

$$\frac{d\mathcal{E}}{d\tau} \leqslant 2\Lambda \int_0^{2\pi} e^{-\tau}\Phi\widetilde{k}d\theta + \sqrt{2\omega}\int_0^{2\pi} e^{-\tau}\Phi\widetilde{k}\lambda\left(\Psi_{\theta\theta} + \Psi\right)d\theta$$

$$\leqslant 3\Lambda \int_0^{2\pi} e^{-\tau}\Phi\widetilde{k}d\theta \ .$$

By (3.18),

$$\mathcal{E}\left(\widetilde{\gamma}(\cdot,\tau)\right) - \mathcal{E}\left(\widetilde{\gamma}(\cdot,t)\right) \leqslant 3\Lambda C_1 \ .$$

We have shown that $\mathcal{E}\left(\widetilde{\gamma}(\cdot,\tau)\right)$ is uniformly bounded in $[0,\infty)$.

As before, the bound on the entropy yields upper bounds for the diameter and length of the normalized flow, and also a positive lower bound for its inradius. Moreover, the normalized curvature and its gradient are also uniformly bounded.

It remains to obtain a positive lower bound for the normalized curvature. We shall do this by representing the flow as polar graphs. To this end, we first observe that the curvature upper bound controls the speed of the flow. Therefore, there exists $\rho > 0$ such that, for all τ_0, we can find a disk $D_\rho\left((x_0, y_0)\right)$ enclosed by $\widetilde{\gamma}(\cdot,\tau)$ for all $\tau \in [\tau_0, \tau_0 + \rho]$. Using (x_0, y_0) as the origin, we represent $\widetilde{\gamma}(\cdot,\tau)$ as

$$\widetilde{\gamma}(\cdot,\tau) = r(\alpha,\tau)(\cos\alpha, \sin\alpha) \ , \quad \alpha \in [0, 2\pi) \ . \tag{3.21}$$

The tangent and normal of $\widetilde{\gamma}$ are given, respectively, by

$$\boldsymbol{t} = \left(r_\alpha\cos\alpha - r\sin\alpha \ , \ r_\alpha\sin\alpha + r\cos\alpha\right)/D \ ,$$

and

$$\boldsymbol{n} = -\left(r_\alpha\sin\alpha + r\cos\alpha \ , \ -r_\alpha\cos\alpha + r\sin\alpha\right)/D \ ,$$

where $D = \left(r_\alpha^2 + r^2\right)^{1/2}$. By differentiating (3.21), we have

$$\frac{\partial \tilde{\gamma}}{\partial \tau} = \frac{\partial \alpha}{\partial \tau} Dt + \frac{\partial r}{\partial \tau}\left(\cos \alpha\,,\ \sin \alpha\right)\,.$$

Therefore, r satisfies

$$\frac{\partial r}{\partial \tau} = r - \frac{D}{r}\left(\Phi \tilde{k} + \sqrt{2\omega}\lambda e^{-\tau}\Psi\right)\,. \tag{3.22}$$

Notice that we also have

$$\sin \theta = \left(r_\alpha \sin \alpha + r \cos \alpha\right)/D\,,$$

and

$$\tilde{k} = \frac{1}{D^3}\left[-rr_{\alpha\alpha} + 2r_\alpha^2 + r^2\right]\,.$$

By differentiating (3.22), we obtain the equation for $\tilde{k} = \tilde{k}(\alpha, \tau)$:

$$\frac{\partial \tilde{k}}{\partial \tau} = \frac{\Phi}{D}\frac{\partial}{\partial \alpha}\left(\frac{1}{D}\frac{\partial \tilde{k}}{\partial \alpha}\right) + \left(\frac{3\Phi_\theta \tilde{k}}{D} - \frac{r_\alpha\left(\Phi \tilde{k} + \sqrt{2\omega}e^{-\tau}\lambda\Psi\right)}{rD}\right.$$

$$\times \left.\frac{\sqrt{2\omega}e^{-\tau}\lambda\Psi_\theta}{D}\right)\frac{\partial \tilde{k}}{\partial \alpha} + \tilde{k}^2\left[\left(\Phi_{\theta\theta} + \Phi\right)\tilde{k} + \sqrt{2\omega}e^{-\tau}\lambda\left(\Psi_{\theta\theta} + \Psi\right)\right] - \tilde{k}\,.$$

During the time interval $[\tau_0, \tau_0 + \rho]$, r is bounded from above and below by positive constants. Hence, this equation is a uniformly parabolic equation with bounded coefficients for $\tilde{k}$. By the Krylov-Safonov's Harnack inequality, or Moser's Harnack inequality [88] (notice that it can be written in divergence form), we conclude that $\tilde{k} \geqslant \delta > 0$ for some δ in $[\tau_0 + \rho/2, \tau_0 + \rho]$. So $\tilde{k}$ has a uniform positive lower bound in $[1, \infty)$.

With two-sided bounds on $\tilde{k}$, we can easily finish the proof of the subconvergence of the normalized flow as follows. First of all, we may assume

$$\Phi\tilde{k} + \sqrt{2\omega}e^{-\tau}\lambda\Psi \geqslant k_0 > 0\,,\ \text{for all } \tau > 0\,. \tag{3.23}$$

Henceforth, the flow is contracting and the support function $\widetilde{h}$ is positive if we fix the shrinking point in the origin. We claim that, in fact, $\widetilde{h} \geqslant k_0/2$ for all τ. For, if $\widetilde{h} < k_0/2$ at some (θ_0, τ_0), by (3.15) and (3.23), $\widetilde{h}(\theta_0, \tau_0 + 1) < 0$, which is impossible.

Now, consider the functional

$$\mathcal{I}(\tau) = \int_0^{2\pi} \left[\left(\frac{\partial \widetilde{h}}{\partial \theta} \right)^2 - \widetilde{h}^2 + 2\Phi \log \widetilde{h} + 2\Lambda e^{-\tau} \right] d\theta ,$$

where Λ is to be chosen later. We have

$$
\begin{aligned}
\frac{d\mathcal{I}}{d\tau} &= -2 \int \frac{1}{\widetilde{kh}} \left(\Phi \widetilde{k} - \widetilde{h} \right) \left(\Phi \widetilde{k} + \sqrt{2\omega} e^{-\tau} \lambda \Psi - \widetilde{h} \right) \\
&\quad -2 \int_0^{2\pi} \Lambda e^{-\tau} d\theta \\
&\leqslant -\int_0^{2\pi} \frac{1}{\widetilde{kh}} \left(\Phi \widetilde{k} - \widetilde{h} \right)^2 + \int_0^{2\pi} \frac{2\omega}{\widetilde{kh}} e^{-2\tau} \lambda^2 \Psi^2 d\theta \\
&\quad -2 \int_0^{2\pi} \Lambda e^{-\tau} d\theta \\
&\leqslant -\int_0^{2\pi} \frac{1}{\widetilde{kh}} \left(\left(\Phi \widetilde{k} - \widetilde{h} \right)^2 + 4\pi \frac{2\omega}{k_0^2} e^{-\tau} \lambda^2 \left| \Psi \right|_{\max}^2 - \Lambda \right) e^{-\tau} \\
&\leqslant 0 ,
\end{aligned}
$$

if we take $\Lambda = 2\omega/k_0^2 \lambda^2 \left| \Psi \right|_{\max}^2$. Hence,

$$\frac{d\mathcal{I}}{d\tau} \leqslant -\int_0^{2\pi} \frac{1}{\widetilde{kh}} \left(\Phi \widetilde{k} - \widetilde{h} \right)^2 \leqslant 0 \qquad (3.24)$$

along the normalized flow. From parabolic regularity theory, there is a uniform Hölder bound on $\widetilde{k}$. From the boundedness of $\mathcal{I}$ and (3.24), we deduce

$$0 = \lim_{\tau \to \infty} \frac{d\mathcal{I}}{d\tau} \left(\widetilde{\gamma}(\cdot, \tau) \right) ,$$

and

$$\lim_{\tau \to \infty} \int_0^{2\pi} \left(\Phi\tilde{k} - \tilde{h}\right)^2 (\cdot, \tau) d\theta = 0 \ .$$

We conclude that there exists a subsequence $\{\tilde{\gamma}(\cdot, \tau_j)\}$, $\tau_j \longrightarrow \infty$, converging smoothly to a self-similar solution of (3.10).

Remark 3.14 The subconvergence of the normalized flow of (3.9) in the contracting case holds without the positivity of Ψ. Besides, in the next chapter, we shall show that embedded self-similar solutions of (3.10) are unique up to homothety when Φ satisfies $\Phi(\theta + \pi) = \Phi(\theta)$. Consequently, the normalized flow converges smoothly to a self-similar solution in this case.

3.5 The stationary case of the ACEF

In this section, we let $\gamma(\cdot, t)$ be a solution of $(3.9)_\lambda$ for $\lambda \in [\lambda_*, \lambda^*]$. We shall first show that its curvature is uniformly bounded from below and above by positive constants.

According to the definition of λ_* and λ^*, we know that the length of $\gamma(\cdot, t)$, $L(t)$, has a positive lower bound. We claim that it admits a uniform upper bound too. In fact, let's compute

$$\frac{d}{dt}\left(\frac{L^2}{A}\right)$$

$$= \frac{L}{A^2}\left(-2A\int_0^{2\pi} \Phi k d\theta - 2A\int_0^{2\pi} \lambda\Psi d\theta + L\int_0^{2\pi} \Phi d\theta + L\int_\gamma \Psi ds\right)$$

$$\leqslant \frac{L}{A^2}\left(\alpha A - \beta L^2\right)$$

$$\leqslant \frac{\beta L}{A}\left(\frac{\alpha}{\beta} - \frac{L^2}{A}\right),$$

for some positive α and β. Hence,

$$\frac{L^2}{A}(t) \leqslant \max\left\{\frac{\alpha}{\beta}, \frac{L^2}{A}(0)\right\} .$$

If L becomes very large, this estimate implies that we can find a disk inside the curve. However, this is impossible since $\lambda \geqslant \lambda_*$. So L must be uniformly bounded from above.

Notice that the isoperimetric estimate also implies a positive lower bound for the inradius of $\gamma(\cdot, t)$ for all t.

Lemma 3.15 *The curvature is uniformly bounded between two positive constants. Moreover, its derivatives are also uniformly bounded for all t.*

Proof: Let h be the support function of a closed, convex curve γ. We define its "support center" (Chou-Wang [32]) to be

$$c = \int_0^{2\pi} h(\theta)x d\theta , \quad x = (\cos\theta, \sin\theta) .$$

It is easy to see that

$$R_{rin} \geqslant h(\theta) - c \cdot x \geqslant \rho r_{in} ,$$

where r_{in} is the inradius of γ and ρ is a positive absolute constant.

Let $c(t)$ be the support center of $\gamma(\cdot, t)$. Consider the auxiliary function

$$w(\theta, t) = \frac{-\partial h/\partial t(\theta, t)}{h(\theta, t) - c(t) \cdot x - \rho r_0/2} ,$$

where $r_{in} \geqslant r_0 > 0$. Suppose the maximum of w over $[0, 2\pi] \times [0, T]$ is attained at some point (θ_0, t_0), $t_0 > 0$. At this point, we have

$$0 = \left(\widehat{h} - \delta\right)^2 w_\theta = -\left(\widehat{h} - \delta\right)\frac{\partial^2 h}{\partial t \partial \theta} + \frac{\partial h}{\partial t}\left[\frac{\partial h}{\partial \theta} - c \cdot \left(-\sin\theta, \cos\theta\right)\right] ,$$

$$0 \leqslant \left(\widehat{h} - \delta\right)^2 w_t = -\left(\widehat{h} - \delta\right)\frac{\partial^2 h}{\partial t^2} + \frac{\partial h}{\partial t}\left(\frac{\partial h}{\partial t} - \frac{dc}{dt} \cdot x\right) ,$$

and

$$0 \geqslant \left(\widehat{h} - \delta\right)^2 w_{\theta\theta} = -\left(\widehat{h} - \delta\right)\frac{\partial^3 h}{\partial t \partial \theta^2} + \frac{\partial h}{\partial t}\left(\frac{\partial^2 h}{\partial \theta^2} + c \cdot x\right),$$

and $\widehat{h} = h - c \cdot x$ and $\delta = \rho r_0/2$. Using the equation

$$h_t = -\left(\Phi k + \lambda \Psi\right),$$

we have

$$\left(\frac{\partial h}{\partial t}\right)^2 \geqslant \frac{\partial h}{\partial t}\frac{dc}{dt} \cdot x + \frac{\partial^2 h}{\partial t^2}\left(\widehat{h} - \delta\right)$$

$$\geqslant \frac{\partial h}{\partial t}\frac{dc}{dt} \cdot x + \Phi k^2\left(\frac{\partial^3 h}{\partial t \partial \theta^2} + \frac{\partial h}{\partial t}\right)\left(\widehat{h} - \delta\right)$$

$$\geqslant \Phi k^2\left(\frac{\partial^2 \widehat{h}}{\partial \theta^2} + \widehat{h} - \delta\right)\frac{\partial h}{\partial t} + \frac{\partial h}{\partial t}\frac{dc}{dt} \cdot x.$$

In other words,

$$w^2 + \frac{\Phi k w}{\widehat{h} - \delta} \geqslant \frac{\Phi k^2 \delta}{\widehat{h} - \delta}w + \frac{dc/dt \cdot x}{\widehat{h} - \delta}(-w).$$

Using $\rho r_0 \leqslant \widehat{h} \leqslant C_0$, C_0 some constant, and

$$\left|\frac{dc}{dt}\right| \leqslant C_1(1 + k),$$

we conclude that w, and, hence, k, are uniformly bounded in $[0, \infty)$.

This upper bound on the curvature yields an upper bound on the speed of the support center. Using $c(t)$ as the origin, there exists positive $\triangle$ and d_0 independent of t such that dist $\left(c(t), \gamma(\cdot, t')\right) \geqslant d_0$ for all t' in $[t, t+\triangle]$. Thus, we may represent $\gamma(\cdot, t')$ as polar graphs and argue as in the previous section that k is uniformly bounded from below by a positive number, and also that there are uniform bounds on the derivatives of k. The proof of Lemma 3.15 is completed. $\square$

The bounds we have obtained so far are not sufficient for convergence. In fact, $\gamma(\cdot, t)$ may move to infinity in constant speed. We shall show that convergence holds if and only if there is some function ξ on S^1 satisfying

$$\Psi\left(\frac{d^2\xi}{d\theta^2} + \xi\right) = \Phi .$$
(3.25)

We modify the flow as follows. Let

$$D = \left\{ (c_1, c_2) : c = |c|(\cos\theta, \sin\theta) , \ 0 \leqslant |c| < -\lambda\Psi(\theta) \right\}$$

and consider the map from D to $\mathbb{R}^2$ given by

$$c \longmapsto \int_0^{2\pi} \frac{\Phi(\theta)e^{i\theta}d\theta}{(c_1\cos\theta + c_2\sin\theta) + \lambda\Psi(\theta)} .$$

It is readily verified that this map is a diffeomorphism onto $\mathbb{R}^2$. Consequently, there exists a unique point c^* in D satisfying

$$(0,0) = \int_0^{2\pi} \frac{\Phi e^{i\theta}d\theta}{c_1^*\cos\theta + c_2^*\sin\theta + \lambda\Psi} .$$
(3.26)

Let h be the support function of $\gamma(\cdot, t)$. We shift it to

$$\widehat{h} = h - \langle c^*, (\cos\theta, \sin\theta)\rangle t .$$

Then,

$$\frac{\partial\widehat{h}}{\partial t} = -\left(\Phi\widehat{k} + \widehat{\Psi}\right) ,$$

where

$$\widehat{\Psi} = \lambda\Psi + \langle c^*, (\cos\theta, \sin\theta)\rangle ,$$

and $\widehat{k} = k$ is the curvature of $\widehat{\gamma}$, the convex curve determined by $\widehat{h}$.

Consider the function $\widehat{\mathcal{I}}$ of $\widehat{h}(\theta, t)$,

$$\widehat{\mathcal{I}}(t) = \int_0^{2\pi} \frac{\Phi(\theta)\widehat{h}(\theta, t)d\theta}{\widehat{\Psi}(\theta)} - \frac{1}{2}\int_0^{2\pi} \left(\widehat{h}_\theta^2 - \widehat{h}^2\right)d\theta .$$

By (3.26), the first term in $\widehat{\mathcal{I}}$ is independent of the choice of the origin. Hence, it is bounded by a constant multiple of the diameter of $\widehat{\gamma}$, whose uniform boundedness has been established. As for the second term in $\widehat{\mathcal{I}}$, we have

$$-\frac{1}{2}\int_0^{2\pi}\left(\widehat{h}_\theta^2 - \widehat{h}^2\right)d\theta = \frac{1}{2}\int_0^{2\pi}\frac{\widehat{h}}{\widehat{k}}d\theta \ ,$$

which is just the area enclosed by $\widehat{\gamma}$. So $\widehat{\mathcal{I}}$ is uniformly bounded for all t. Now,

$$\frac{d\widehat{\mathcal{I}}}{dt} = -\int \frac{(\Phi\widehat{k}+\widehat{\Psi})^2}{\widehat{k}\widehat{\Phi}}d\theta \leqslant 0 \ .$$

By Lemma 3.15,

$$\sup_\theta\left|\widehat{h}_t(\theta,t)\right| \longrightarrow 0 \quad \text{as} \quad t\longrightarrow\infty \ .$$

To furnish the last step in proving convergence, we introduce the "modified support center,"

$$\widehat{c}(t) = \int_0^{2\pi}\frac{\Phi}{\widehat{\Psi}^2}\widehat{h}(\theta,t)(\cos\theta,\sin\theta)d\theta \ ,$$

of $\widehat{\gamma}$. Notice that $\widehat{c}(t)$ may not lie inside $\widehat{\gamma}(\cdot,t)$. We claim: there exists $\widehat{c}(\infty) \in \mathbb{R}^2$ such that

$$\lim_{t\to\infty}\widehat{c}(t) = \widehat{c}(\infty) \ .$$

Because $\widehat{k}$ is the curvature of a closed curve and (3.26),

$$\begin{aligned}
(0,0) &= \int_0^{2\pi}\frac{e^{i\theta}}{\widehat{k}(\theta,t)}d\theta \\
&= -\int_0^{2\pi}\frac{\Phi e^{i\theta}}{\widehat{\Psi}(1+\widehat{h}_t\widehat{\Psi}^{-1})}d\theta \\
&= -\int_0^{2\pi}\frac{\Phi}{\widehat{\Psi}}e^{i\theta}d\theta + \int_0^{2\pi}\frac{\Phi}{\widehat{\Psi}^2}e^{i\theta}\widehat{h}_td\theta + O\left(\widehat{h}_t^2\right) \ , \\
&= \int_0^{2\pi}\frac{\Phi}{\widehat{\Psi}^2}e^{i\theta}\widehat{h}_td\theta + O\left(\widehat{h}_t^2\right) \ ,
\end{aligned}$$

for small $\widehat{h}_t$, it follows that

$$\left|\frac{d\widehat{c}}{dt}\right| \leqslant C\int_0^{2\pi} \widehat{h}_t^2 \, d\theta$$

$$\leqslant C\left|\frac{d\widehat{\mathcal{I}}}{dt}\right| .$$

Consequently,

$$\left|\widehat{c}(t) - \widehat{c}(t')\right| \leqslant C\left|\widehat{\mathcal{I}}(t) - \widehat{\mathcal{I}}(t')\right| \longrightarrow 0$$

as $t, t' \longrightarrow \infty$.

Now we can show that $\widehat{\gamma}(\cdot, t)$ cannot escape from the plane. For each t, write

$$\widehat{h} = h' + \langle \ell, (\cos\theta, \sin\theta) \rangle ,$$

where h' is positive and $\ell \in \mathbb{R}^2$. It suffices to show that $\{\ell = \ell(t)\}$ are uniformly bounded. After rotating the axes, one may assume $\ell_2 = 0$. Then,

$$\begin{aligned}
\widehat{c}_1(t) &= \int_0^{2\pi} \frac{\Phi}{\widehat{\Psi}^2}\left(h' + \langle \ell, (\cos\theta, \sin\theta) \rangle \right)\cos\theta \, d\theta \\
&= \int_0^{2\pi} \frac{\Phi}{\widehat{\Psi}^2} h' \cos\theta \, d\theta + |\ell| \int_0^{2\pi} \frac{\Phi}{\widehat{\Psi}^2}\cos^2\theta \, d\theta .
\end{aligned}$$

Therefore,

$$\begin{aligned}
|\ell| &\leqslant C\left(\left|\widehat{c}_1(t)\right| + L(t)\right) \\
&\leqslant C\left(1 + \left|\widehat{c}_1(\infty)\right| + L_0\right) ,
\end{aligned}$$

for all large t.

Now, by the Blaschke Selection Theorem, any sequence $\{\widehat{h}(\cdot, t_j)\}, t_j$

$\to \infty$, contains a subsequence which converges smoothly to a stationary solution. To show that the subconvergence is in fact a uniform convergence, it suffices to show that all limit curves are identical.

Recall that all stationary solutions are unique up to translations. Suppose $\widehat{\gamma}(\cdot, t_j)$ and $\widehat{\gamma}(\cdot, t'_j)$ converge to γ_1 and γ_2, respectively. The support functions of γ_1 and γ_2, h_1 and h_2, satisfy

$$h_2 - h_1 = \ell \cdot (\cos \theta, \sin \theta) \ ,$$

for some $\ell \in \mathbb{R}^2$. From

$$\int_0^{2\pi} \frac{\Phi}{\widehat{\Psi}^2} (\ell_1 \cos \theta + \ell_2 \sin \theta)(\cos \theta, \sin \theta) d\theta$$

$$= \int_0^{2\pi} \frac{\Phi}{\widehat{\Psi}^2} (\cos \theta, \sin \theta) h_1 d\theta - \int_0^{2\pi} \frac{\Phi}{\widehat{\Psi}^2} (\cos \theta, \sin \theta) h_2 d\theta$$

$$= \lim_{t_j \to \infty} \widehat{c}(t_j) - \lim_{t_0^1 \to \infty} \widehat{c}(t'_j)$$

$$= (0,0) \ ,$$

we conclude $\ell = (0,0)$.

Finally, we claim that $\lambda_* = \lambda^*$. For, let γ_* and γ^* be the modified flow of $(3.9)_\lambda$ for $\lambda = \lambda_*$ and $\lambda = \lambda^*$, respectively. We have

$$\gamma_*(\infty) \quad = \quad \lim_{t \to \infty} \gamma_*(\cdot, t) \tag{3.27}$$

$$= \quad \lim_{t \to \infty} \gamma^*(\cdot, t)$$

$$= \quad \gamma^*(\infty) \ .$$

On the other hand, $g = h_* - h^*$ satisfies the equation

$$\frac{\partial g}{\partial t} = \Phi k_* k^* \left(\frac{\partial^2 g}{\partial \theta^2} + g \right) + (\lambda^* - \lambda_*) \Psi \ .$$

Were $\lambda^* > \lambda_*$, $\min_\theta (h_* - h^*)(t)$ becomes positive and is nondecreasing for $t > 0$. Hence, $h_*(\cdot, \infty) > h^*(\cdot, \infty)$, contradicting (3.27). So we must have $\lambda_* = \lambda^*$. The proof of Part (ii) in Theorem 3.12 is completed. $\qquad\square$

3.6   The expanding case of the ACEF

According to the definition of λ^*, for all $\lambda < \lambda^*$, the flow $\gamma(\cdot, t)$ of $(3.9)_\lambda$ possesses the following property: for any bounded subset K of the plane, there exists t_K such that K is contained inside $\gamma(\cdot, t)$ for all $t \geqslant t_K$. In this section, we study the asymptotic behaviour of this flow.

Intuitively speaking, in this case, the curvature of the flow is eventually negligible. In view of (3.9), the length of $\gamma(\cdot, t)$ grows linearly. Therefore, we consider the normalization given by

$$\widetilde{\gamma}(\cdot, t) = \gamma(\cdot, t)/t.$$

We shall assume that the polar graph of $1/\Psi$ is uniformly convex. This assumption means that the Wulff region of Ψ, $W(\Psi)$, has a uniformly convex boundary. One can directly verify that this is equivalent to the inequality

$$\Psi_{\theta\theta} + \Psi > 0 \, . \tag{3.28}$$

Theorem 3.12 (iii) is contained in the following proposition.

Proposition 3.16 *Let γ be a solution of $(3.9)_\lambda$, $\lambda < \lambda^*$. Suppose that (3.28) holds. Then*

$$\widetilde{h}(\cdot, t) + \lambda\Psi(\cdot) = O(t^{-1} \log t), \quad t \to \infty,$$

uniformly, where $\widetilde{h}$ is the support function of $\widetilde{\gamma}$. Moreover,

$$-\lambda \left(\Psi_{\theta\theta} + \Psi \right) \widetilde{k}(\cdot, t) \to 1 \quad \text{uniformly and}$$

$$\left\| \frac{d^n}{d\theta^n} \left(\Psi_{\theta\theta} + \Psi \right) \widetilde{k}(\cdot, t) \right\|_{L^\infty} = O(t^{-\alpha}),$$

uniformly for any $n \geqslant 1$ and $\alpha \in (0,1)$ as $t \to \infty$.

We shall show at the end of this section that (3.28) is necessary for C^2-convergence of $\widetilde{h}$.

By introducing the new time scale $\tau = \log t$, the equations for $\widetilde{h}$ and $\widetilde{k}$ are, respectively, given by

$$-\frac{\partial \widetilde{h}}{\partial \tau} = e^{-\tau} \Phi \widetilde{k} + \lambda \Psi + \widetilde{h} \tag{3.29}$$

and

$$\frac{\partial \widetilde{k}}{\partial \tau} = \widetilde{k}^2 \frac{\partial^2}{\partial \theta^2} (e^{-\tau} \Phi \widetilde{k} + \lambda \Psi) + \widetilde{k}^2 (e^{-\tau} \Phi \widetilde{k} + \lambda \Psi) + \widetilde{k} . \tag{3.30}$$

Lemma 3.17 *There exists k_0 which depends on Φ and Ψ only such that $\widetilde{k}$ is uniformly bounded in $[0, 2\pi] \times [0, \infty)$ if*

$$\max_\theta k(\theta, 0) \leqslant k_0 .$$

Proof: From (1.17), we have

$$\frac{dk_{\max}}{dt} \leqslant \max_\theta \left\{ (\Phi_{\theta\theta} + \Phi) k_{\max}(t) + \lambda \Psi_{\theta\theta} + \lambda \Psi \right\} k_{\max}^2 .$$

Hence, if

$$\max_\theta \left\{ (\Phi_{\theta\theta} + \Phi) k_{\max}(t) + \lambda \Psi_{\theta\theta} + \lambda \Psi \right\} < 0 ,$$

$k_{\max}$ is decreasing in t and

$$\frac{dk_{\max}}{dt} \leqslant -C_0 k_{\max}^2(t) ,$$

which means

$$k_{\max}(t) \leqslant \frac{k_{\max}(0)}{1 + C_0 k_{\max}(0)t} \, , \qquad t \in [0, \infty) \, .$$

$\square$

Lemma 3.18 *Let $\widetilde{h}_C$ be the support function of the flow $(3.9)_\lambda$, $\lambda < \lambda^*$, whose initial curve is a large circle. Then*

$$\widetilde{h}_C(\cdot, t) + \lambda \Psi(\cdot) \to 0$$

uniformly as $t \to \infty$.

Proof: We integrate (3.29) to get

$$e^\tau (\widetilde{h}_C(\theta, \tau) + \lambda \Psi(\theta)) = \widetilde{h}_C(\theta, 0) + \lambda \Psi(\theta) - \int_0^\tau \Phi(\theta) \widetilde{k}(\theta, s) ds \, .$$

When the initial circle is so large that

$$\max_\theta \left\{ (\Phi_{\theta\theta} + \Phi) k_{\max}(0) + \lambda \frac{d^2 \Psi(\theta)}{d\theta^2} + \lambda \Psi(\theta) \right\} < 0,$$

it follows from Lemma 3.17 that $\widetilde{k}$ is uniformly bounded. Consequently, we have

$$\widetilde{h}_C(\cdot, \tau) + \lambda \Psi(\cdot) = O(\tau e^{-\tau})$$

uniformly as $\tau \to \infty$.

$\square$

Lemma 3.19 $\widetilde{h}(\cdot, \tau) + \lambda \Psi(\cdot)$ *tends to zero uniformly as $\tau \to \infty$.*

Proof: Since $\gamma(\cdot, t)$ expands, we may assume without loss of generality that $\gamma(\cdot, t)$ is pinched between two large circles. Then the desired result follows from Lemma 3.18.

$\square$

Lemma 3.20 $\widetilde{k}$ *is uniformly bounded in $[0, 2\pi] \times [0, \infty)$.*

Proof: First, we claim that, for $\varepsilon > 0$, there exists τ^* such that

$$\max_\theta k(\theta, \tau^*) \leqslant \varepsilon.$$

For, given $\varepsilon > 0$, we divide $[0, 2\pi]$ into approximately $[1/\varepsilon]$-many subintervals I_js of length equal to ε. For $\tau > 0$, we have

$$\int_\tau^{\tau+1} \Phi(\theta)\widetilde{k}(\theta, s)ds = -(\lambda\Psi + \widetilde{h})(\theta, \tau + 1)e^{\tau+1} + (\lambda\Psi + \widetilde{h})(\theta, \tau)e^\tau$$

and

$$\int_\tau^{\tau+1} \fint_{I_j} \Phi(\theta)\widetilde{k}(\theta, s)d\theta ds$$

$$= -e^{\tau+1}\fint_{I_j}(\lambda\Psi + \widetilde{h})(\theta, \tau + 1)d\theta + e^\tau\fint_{I_j}(\lambda\Psi + \widetilde{h})(\theta, \tau)d\theta,$$

where $\fint_{I_j}$ is the average over I_j. Summing over I_js,

$$\int_\tau^{\tau+1}\left(\sum_j \fint_{I_j} \Phi(\theta)\widetilde{k}(\theta, s)d\theta + \int_0^{2\pi} \Phi(\theta)\widetilde{k}(\theta, s)d\theta\right)ds$$

$$= -e^{\tau+1}\left[\sum_j \fint_{I_j}(\lambda\Psi + \widetilde{h})(\theta, \tau + 1)d\theta + \int_0^{2\pi}(\lambda\Psi + \widetilde{h})(\theta, \tau + 1)d\theta\right]$$

$$+e^\tau\left[\sum_j \fint_{I_j}(\lambda\Psi + \widetilde{h})(\theta, \tau)d\theta + \int_0^{2\pi}(\lambda\Psi + \widetilde{h})(\theta, \tau)d\theta\right].$$

By the mean-value theorem, there exists $\tau^* \in (\tau, \tau + 1)$ such that

$$\sum_j \fint_{I_j} \Phi(\theta)k(\theta, \tau^*)d\theta + \int_0^{2\pi} \Phi(\theta)k(\theta, \tau^*)d\theta$$

$$\leqslant 2e\left(1 + \frac{1}{\varepsilon}\right)\left[\int_0^{2\pi} |\lambda\Psi + \widetilde{h}|(\theta, \tau)d\theta + \int_0^{2\pi} |\lambda\Psi + \widetilde{h}|(\theta, \tau + 1)d\theta\right.$$

$$\left. + \max_\theta\{|\lambda\Psi + \widetilde{h}|(\theta, \tau) + |\lambda\Psi + \widetilde{h}|(\theta, \tau + 1)\}\right].$$

Since by Lemma 3.19, $\lambda\Psi+\widetilde{h}$ tends to zero uniformly, for sufficiently large τ we have

$$\sum_j \fint_{I_j} \Phi(\theta)k(\theta,\tau^*)d\theta + \int_0^{2\pi} \Phi(\theta)k(\theta,\tau^*)d\theta < \varepsilon.$$

Suppose that $\max_\theta k(\theta,\tau^*)$ is attained at $\theta=\theta_0$, θ_0 in some I_j. We have

$$\left| \max_\theta k(\theta,\tau^*) - \fint_{I_j} k(\theta,\tau^*)d\theta \right|$$

$$\leqslant \left\| \frac{\partial k}{\partial\theta}(\cdot,\tau^*) \right\|_{L^\infty} \varepsilon$$

$$\leqslant C\varepsilon$$

by Lemma 3.13. Therefore,

$$\max_\theta k(\theta,\tau^*) \leqslant (1+C)\varepsilon.$$

Now, we can apply Lemma 3.17 to γ, using $\tau=\tau^*$ as the initial time, to show that $\widetilde{k}$ is uniformly bounded. $\square$

To show higher order convergence, we look at the equation satisfied by

$$w = -\lambda\left(\Psi_{\theta\theta}+\Psi\right)\widetilde{k}.$$

By a direct computation

$$\frac{\partial w}{\partial\tau} = e^{-\tau}\widetilde{k}^2\Phi\frac{\partial^2 w}{\partial\theta^2} + e^{-\tau}B\frac{\partial w}{\partial\theta} + e^{-\tau}Cw + w(1-w)\,, \quad (3.31)$$

where

$$B = 2\widetilde{k}^2\left[\Phi_\theta + \frac{(\Psi_{\theta\theta}+\Psi)_\theta}{-(\Psi_{\theta\theta}+\Psi)}\Phi\right],$$

$$C = \widetilde{k}^2\left[\frac{(\Psi_{\theta\theta}+\Psi)_\theta\Phi_\theta}{-(\Psi_{\theta\theta}+\Psi)} + \frac{2(\Psi_{\theta\theta}+\Psi)_\theta^2\Phi}{(\Psi_{\theta\theta}+\Psi)^2} + \frac{(\Psi_{\theta\theta}+\Psi)_{\theta\theta}\Phi}{-(\Psi_{\theta\theta}+\Psi)}\right.$$

$$\left. + \frac{(\Phi_{\theta\theta}+\Phi)w^2}{\lambda^2(\Psi_{\theta\theta}+\Psi)^2}\right].$$

Therefore, by Lemma 3.20 we have

$$\frac{dw_{\max}}{d\tau} \leqslant Ae^{-\tau} + w_{\max}(1 - w_{\max}), \quad \text{and}$$

$$\frac{dw_{\min}}{d\tau} \geqslant -Ae^{-\tau} + w_{\min}(1 - w_{\min})$$

for some constant A. It follows that w tends to 1 uniformly as τ approaches ∞.

By differentiating (3.31), we see that $u = \partial w/\partial\theta$ satisfies the equation

$$\frac{\partial u}{\partial \tau} = e^{-\tau}\left(A'\frac{\partial^2 u}{\partial\theta^2} + B'\frac{\partial u}{\partial\theta} + C'u\right) + e^{-\tau}\frac{\partial C}{\partial\theta}w + (1 - 2w)u$$

where A', B', and C' are uniformly bounded in $[0, 2\pi] \times [0, \infty)$. Since w tends to 1 uniformly, it is easy to see that, for any $\alpha \in (0, 1)$,

$$\frac{\partial w}{\partial\theta}(\cdot, \tau) = O(e^{-\alpha\tau}) \quad \text{uniformly as } \tau \to \infty.$$

Higher order estimates can be obtained in the same way. The proof of Proposition 3.16 is completed.

Finally, we want to show that (3.28) is necessary for C^2-convergence of $\tilde{h}$. More precisely, we have,

Proposition 3.21 *If $\tilde{h}(\cdot, t)$ converges to some $\tilde{h}$ in C^2-norm, then (3.28) holds.*

Proof: From (3.30),

$$\frac{d\tilde{k}_{\min}}{d\tau} \geqslant \tilde{k}_{\min}\left[1 + \tilde{k}_{\min}\min_{\theta}\lambda(\Psi_{\theta\theta} + \Psi) + e^{-\tau}\tilde{k}_{\min}^2\min_{\theta}(\Phi_{\theta\theta} + \Phi)\right],$$

we can see that $\tilde{k}_{\min}$ has a positive lower bound. When $\tilde{h}(\cdot, t)$ tends to $\tilde{h}$ in C^2-norm, the curve determined by $\tilde{h}$ must have positive curvature. However, by Lemma 3.19, $\tilde{h}$ is equal to $-\lambda\Psi$. Hence, (3.28) holds. $\square$

Notes

Gage-Hamilton Theorem. Theorems 3.4 and 3.5 were proved in [58]. The notion of entropy was implicitly introduced and its monotonicity property was established in this paper. To show the convergence to a circle, they used an inequality of Gage [54] (see §4.4) which asserts that the isoperimetric ratio decreases along the flow for convex curves. In another proof [72], Hamilton appeals to the characterization of the self-similar solution due to Abresh and Langer. Finally, Andrews [10] deduced the convergence from a characterization of equality in the entropy inequality which follows from the Minkowski inequality. Our proof here is elementary and is based on the monotonicity of the functional $\mathcal{F}$ (Firey [51]) and Proposition 2.3.

The models. The CSF was posed as a phenomenological model for the dynamics of grain boundaries in Mullins [?] in the fifties. Since then, it and the more general ACEF have arisen in many different areas, including phase transition, crystal growth, flame propagation, chemical reaction, and mathematical biology. We briefly discuss three of these areas.

First, in the sharp interface approach to the theory of phase transition, different bulk phases are separated by a sharp interface which is assumed to be a plane curve. The temperatures in both bulks satisfy the heat equation and the interface is a free boundary. However, in the case of a perfect conductor, the temperatures are constant in both phases and it is possible to reduce the motion of the interface to a single equation. This was done in Gurtin [71]. In fact, he first derived a balance law relating the capillary force of the interface and the interactive force. Next, he used a version of the second law of thermodynamics to obtain constitutive restrictions on

the form of the equation. The resulting law of motion of the interface
is

$$b(\theta)\frac{\partial\gamma}{\partial t} = \big((f_{\theta\theta} + f)k + F\big)\boldsymbol{n} \ ,$$

where F is a constant (the difference in energy between the phases),
$b > 0$ a material function that characterizes the kinetics, and f the
interfacial energy. The dependence of b and f on the normal angle θ
reflects anisotropy. In case $f_{\theta\theta} + f > 0$, it was proved in Angenent-
Gurtin [17] that (a) if $F \geqslant 0$, then $\omega < \infty$ and the enclosed area
$A(t) \to 0$ as $t \uparrow \omega$; and (b) if $F < 0$, then (i) if the initial length is
sufficiently small, then $\omega < \infty$ and $A(t) \to 0$ and (ii) if the initial
area is sufficient large, then $\omega = \infty$, $A(t) \to \infty$ and the isoperimetric ratio remains bounded. They also conjectured that in case
(b)(ii), the flow is asymptotic to a Wulff region of $1/b(\theta)$. This con-
jecture was subsequently confirmed by Soner [100]. Our Theorem
3.12 (Chou-Zhu [35]) gives a precise and complete description of the
flow for convex curves. We point out that the flow may develop self-
intersections when it is non-convex. In this case, one has to use the
level-set approach ([100]). When $\Psi \equiv 0$, the flow was also studied
in Gage [57] and Gage-Li [59] in the context of Minkowski geometry.
In particular, Theorem 3.12(i) was proved in [59] for this case.

One should keep in mind that the model also makes sense when
$f_{\theta\theta} + f \leqslant 0$ or when it is not C^2. As a special important case, we
comment on the crystalline energy which, by definition, has a polygo-
nal Frank diagram (the Frank diagram is the polar graph of $1/f(\theta)$).
Its derivatives have jumps at the vertices. The motion laws for the
crystalline energy were derived by Angenent-Gurtin [17] and Taylor
[106] independently. This is a system of ODEs on the sides of the
polygon under evolution. Crystalline versions of the Gage-Hamilton
theorem and the Grayson convexity theorem can be found in Stancu

[103], [104] and Giga-Giga [62] respectively. For a level-set approach, see also Giga-Giga [61].

Second, in the field equation approach to phase transition, one considers reaction-diffusion equations such as the Allen-Cahn equation,

$$u_t = \Delta u - W'(u) \ , \quad (x,t) \in \mathbb{R}^n \times (0, \infty) \ , \qquad (3.32)$$

where W is a double-well potential having exactly two strict minima, say, at $u = -1$ and 1. The states $u = \pm 1$ represent stable phases. It is well-known that (3.32) admits a unique travelling wave solution $\phi(x_1 - ct, x_2, \cdots, x_n)$, satisfying $\phi(\pm\infty) = \pm 1$ and $\phi' > 0$ where the wave speed is given by

$$c = \left(\int_{-\infty}^{\infty} \phi'^2 \right)^{-1} (W(-1) - W(1)) \ .$$

When $c = 0$, the CSF arises as the singular limit of a general solution of (3.32) as $t \to \infty$. In fact, let $u^\varepsilon(x,t) = u(x/\varepsilon, t/\varepsilon^2)$. Then u^ε satisfies

$$u_t^\varepsilon = \Delta u^\varepsilon - \varepsilon^{-2} W'(u^\varepsilon) \ . \qquad (3.33)$$

Let's focus on $n = 2$ and let $\gamma_0 = \{u_0(x) = 0\}$ and $D_0 = \{u_0(x) > 0\}$ for a given initial u_0. In DeMottoni-Schatzman [40] and Chen [27], the following result is proved:

Theorem. *Let u^ε be the solution of (3.33) satisfying $u^\varepsilon(x,0) = u_0$. Let $\gamma(\cdot, t)$ be the CSF starting at γ_0 and D_t the region it encloses. Then as $\varepsilon \to 0$,*

$$u^\varepsilon \longrightarrow \begin{cases} 1 & \bigcup_{t>0} (D_t \times \{t\}) \\ -1 & \bigcup (\mathbb{R}^2 \setminus \overline{D}_t \times \{t\}) \end{cases}$$

compactly in $\mathbb{R}$.

So the interface moves approximately under the CSF when time is long enough. Incidentally, we mention that, when the wave speed is non zero, one should take the scaling $u^\varepsilon(x,t) = u(x/\varepsilon, t/\varepsilon)$ and the interface moves approximately under the eikonal equation $\gamma_t = c\boldsymbol{n}$. Higher dimensional results of this kind can be found in Souganidis [101] and the references therein.

Finally, the (isotropic) curvature-eikonal equation was derived in the study of wave propagation in weakly excitable media (Keener-Sneyd [84] and Meron [89]). (The name curvature-eikonal equation is taken from [84].) This equation is relevant in a number of biological and chemical contexts, such as the wave front propagation in the excitable Belousov-Zhabotinsky reagent, the calcium waves in *Xenopus* oocytes, and the studies in myocardial tissue. Here the reaction diffusion system is of the form

$$\begin{cases} \varepsilon u_t = \varepsilon^2 \Delta u + f(u,v) \\ v_t = \varepsilon D \Delta v + g(u,v) \ , \end{cases}$$

where u is the activator (propagator) and v is the inhibitor (controller) in the medium. A specific example of the reaction term is the FitzHugh-Nagumo dynamics. When ε is very small, one can formally show that the system can be described by

$$\begin{cases} \gamma_t = (\varepsilon k + c(v))\boldsymbol{n} \\ v_t = \varepsilon D \Delta v + g(u_\pm(v), v) \ , \end{cases}$$

where $u_\pm(v)$ are the two stable roots of $f(u,v) = 0$ (Keener [82]). In general, the first equation in this system describes a free boundary while the second equation should be satisfied by v in the two separating regions. We obtain the curvature-eikonal flow when c is independent of v. In general, one expects this coupled system to be useful in explaining certain commonly observed stable patterns, such as the spirals, in dynamics in excitable media.

The CSF and the curvature-eikonal flow are also relevant in other physical contexts, see, e.g., Deckelnick-Elliott-Richardson [42] for the motion of a superconducting vortex, and Frankel-Sivahinsky [52] for the propagation of flame front.

The level-set approach in curvature flows. This approach provides a unique, globally defined generalized solution for many geometric flows, including the ACEF. Although the results are not only valid for curves but also for many geometric flows, including the anisotropic\isotropic mean curvature flow in higher dimensions, for the purpose of illustration we state them for curves only. To start, let's consider the CSF. Assume that there is a function $u(x,t)$ whose level set $\gamma(\cdot, t) = \{u(x, t) = c\}$ satisfies the CSF for each fixed c and that $\{u(x, t) > c\}$ is the bounded set enclosed by $\gamma(\cdot, t)$. Then we have $\boldsymbol{n} = \nabla u / |\nabla u|$ and $\gamma_t \cdot n = u_t / |\nabla u|$ on $\gamma(\cdot, t)$. Therefore, u satisfies the equation

$$u_t = |\nabla u| \operatorname{div} \left(\frac{\nabla u}{|\nabla u|} \right). \tag{3.34}$$

Conversely, one can verify that, for any solution of (3.34), the set $\gamma(\cdot, t) = \{u(x, t) = c\}$ satisfies the CSF. Consequently, one may use this connection to define a generalized solution of the CSF whenever a solution of (3.34) is given. The equation (3.34) is weakly parabolic and is not defined at $|\nabla u| = 0$. It is crucial that, nevertheless, a notion of generalized solution, namely the viscosity solution, is available for this equation. It turns out this consideration works for other flows. To state a sample result, let's consider (1.2) where F depends on θ and q only. The corresponding equation is

$$u_t = |\nabla u| F \left(\frac{\nabla u}{|\nabla u|}, \operatorname{div} \frac{\nabla u}{|\nabla u|} \right). \tag{3.35}$$

Denote by $C_\alpha(\mathbb{R}^2)$ the family of continuous functions u such that $u - \alpha$ is of compact support. We have (Chen-Gigo-Goto [28]):

Theorem. *Assume that $F = F(\theta, q)$ is uniformly parabolic in (1.2). Let D_0 be an open set and $\gamma_0 \supseteq \partial D_0$ be a bounded set. Then for any function u_0 in $C_\alpha(\mathbb{R}^2), \alpha < 0$, which satisfies $\{u_0 > 0\} = D_0$ and $\{u_0 = 0\} = \gamma_0$, there exists a unique viscosity solution of (3.35) satisfying $u(\cdot, 0) = u_0$. Moreover, the sets $\gamma(\cdot, t) = \{u(x, t) = 0\}$ and $D_t = \{u(x, t) > 0\}$ depend only on γ_0 and D_0 but not on α or u_0.*

Consequently, the flow (1.2) admits a unique generalized solution for all t. Notice that in some cases such as the CSF, $\gamma(\cdot, t)$ becomes empty for $t \geqslant \omega$. A mathematical theory on the level-set approach was introduced in [28] and Evans-Spruck [50] independently. In [50] and its sequels, the mean curvature flow is discussed in some depth, while in [28] other geometric curvature flows are also treated. For further work, we refer to Giga-Goto-Ishii-Sato [63], [100], and the survey [101]. For a geometric measure-theoretic approach to the mean curvature flow, one may consult Brakke [22] and Ilmanen [81]. See also Ambrosio-Soner [7] for another approach by De Giorgi.

Chapter 4

The Convex Generalized Curve Shortening Flow

In this chapter, we study the Cauchy problem for the anisotropic generalized curve shortening flow (AGCSF),

$$\frac{\partial \gamma}{\partial t} = \Phi(\theta) k^\sigma \boldsymbol{n} \ , \ \sigma > 0 \ , \tag{4.1}_\sigma$$

where Φ is a smooth positive, 2π-periodic function of the normal angle and γ_0 is a given uniformly convex, embedded closed curve. According to the results in §3.2, this flow preserves convexity and shrinks to a point in finite time. To examine its ultimate shape, we normalize the flow using the shrinking point as the origin so that its enclosed area is always equal to π. In terms of the support function and curvature,

$$\tilde{h} \ = \ \left(\frac{A(t)}{\pi}\right)^{-1/2} h \ , \qquad \text{and}$$

$$\tilde{k} \ = \ \left(\frac{A(t)}{\pi}\right)^{1/2} k \ ,$$

we have the equations

$$\tilde{h}_\tau = -\Phi \tilde{k}^\sigma + \left(\oint_{S^1} \Phi \tilde{k}^{\sigma-1} d\theta\right) \tilde{h} \ , \tag{4.2}$$

and

$$\tilde{k}_\tau = \tilde{k}^2 \left[\left(\Phi \tilde{k}^\sigma\right)_{\theta\theta} + \Phi \tilde{k}^\sigma\right] - \left(\oint_{S^1} \Phi \tilde{k}^{\sigma-1} d\theta\right) \tilde{k} \ , \tag{4.3}$$

93

where $d\tau/dt = (A(t)/\pi)^{-(1+\sigma)/2}$. It will be shown, (see Lemma 4.6), that $\tau = O(\log(\omega - t))$. Hence, (4.2) and (4.3) are defined in $[\tau_0, \infty)$ for some τ_0. Formally, in case $\tilde{h}_\tau$ tends to zero as $\tau \longrightarrow \infty$, the limit satisfies the equation

$$\Phi k^\sigma = \mu h , \tag{4.4}$$

where μ is a positive constant. In other words, it is a contracting self-similar solution for (4.1).

We shall establish the subconvergence of the normalized flow to self-similar solutions for $\sigma \in (1/3, 1)$ in Section 2. Some useful inequalities from the theory of convex bodies are listed in Section 1. In Section 3, we study the affine curve shortening problem, that is, $\sigma = 1/3$ and $\Phi \equiv 1$ in (4.1). It has an invariant formulation in the context of affine geometry. We shall show that convergence to ellipses holds for the normalized flow. In the last section, we present a uniquence result on self-similar solutions for (4.1) where $\sigma \geqslant 1$ and Φ is symmetric.

4.1 Results from the Brunn-Minkowski Theory

In this section, we collect some basic results from the Brunn-Minkowski Theory of convex bodies. Although the natural setting is convex bodies, for our purpose it is sufficient to state them for convex bodies with a smooth boundary. As a result, they will appear in slightly restrictive form.

Let h and φ be two smooth functions defined on S^1. We shall use the following notations:

$$A_0 = \int_{S^1} h A[h] d\theta ,$$

$$A_1 \;=\; \int_{S^1} \varphi A[\varphi] d\theta \;, \qquad\qquad \text{and}$$

$$A_{01} \;=\; \int_{S^1} \varphi A[h] d\theta = \int_{S^1} h A[\varphi] dd\theta \;,$$

where $A[f] = f_{\theta\theta} + f$ for any function f.

Theorem 4.1 (Minkowski inequality). *We have*

$$A_{01}^2 \geqslant A_0 A_1$$

under either one of the following conditions: (i) $A[h] \geqslant 0$ and $A[\varphi] \geqslant 0$, or (ii) $A[h] > 0$. Moreover, equality in this inequality holds if and only if there exist a positive λ and a point (x_0, y_0) such that

$$\varphi(\theta) = \lambda h(\theta) + \langle (x_0, y_0), (\cos\theta, \sin\theta) \rangle \;,$$

for all $\theta \in S^1$.

Theorem 4.1 under assumption (i) is the standard form of the Minkowski inequality and usually is deduced from the Brunn-Minkowski inequality. Under (ii), we may apply the inequality to the functions h and $\varphi + \lambda h$, where λ has been chosen so large that $A[\varphi + \lambda h] \geqslant 0$.

The following two results are about stability. They provide effective lower bounds of the difference

$$\triangle = A_{01}^2 - A_0 A_1 \;.$$

Let K_0 and K_1 be two convex sets with support functions h and φ, respectively. The circumradius and inradius of K_0 with respect to K_1 are given by

$$R(K_0, K_1) = \inf \left\{ R > 0 : K_0 \subseteq R(K_1 - (x, y)) \text{ for some } (x, y) \right\}$$

and

$$r(K_0, K_1) = \sup \left\{ r > 0 : r\left(K_1 - (x,y)\right) \subseteq K_0 \text{ for some } (x,y) \right\} ,$$

respectively. When K_1 is the unit disk, the circumradius and inradius of K_0 are the usual circumradius and inradius of a convex body (or its boundary). Our first stability estimate is an anisotropic version of the Bonnesen inequality.

Theorem 4.2 *Let $R = R(K_0, K_1)$ and $r = r(K_0, K_1)$. We have*

$$A_0 - 2\rho A_{01} + \rho^2 A_1 \leqslant 0$$

for all $\rho \in [r, R]$. Consequently,

$$\triangle \geqslant \frac{A_1^2}{4}(R - r)^2 .$$

Next, let's denote by $\widehat{h}$ and $\widehat{\varphi}$ the support functions $\widehat{K_0}$ and $\widehat{K_1}$, which are obtained from K_0 and K_1 by translating their centers of mass to the origin and rescaling their perimeters to 1. We have:

Theorem 4.3 *There exists a constant C depending only on the diameter and inradius of K_0 such that*

$$\triangle \geqslant C A_{01}^2 \int_{S^1} (\widehat{h} - \widehat{\varphi})^2 d\theta .$$

Finally, we state two isoperimetric inequalities somehow related to the affine and $SL(2, \mathbb{R})$–geometries:

Theorem 4.4 (The affine isoperimetric inequality). *For any closed embedded convex curve γ, we have*

$$\left(\int_{S^1} k^{-2/3} d\theta \right)^3 \leqslant 8\pi^2 A ,$$

and the equality holds if and only if γ is an ellipse.

As will be explained in Section 3, the integral on the left-hand side of this inequality is the affine perimeter $\mathcal{L}$ of γ. Thus, this inequality asserts that only the ellipses maximize the affine isoperimetric ratio $\mathcal{L}^3/A$.

Theorem 4.5 (Blaschke-Santaló inequality). *For any closed embedded convex curve γ, there exists a point (x_0, y_0) such that*

$$\int_{S^1} \frac{1}{h^2} d\theta \leqslant \frac{2\pi^2}{A} \ ,$$

where h is the support function of γ with respect to (x_0, y_0). Furthermore, equality in this inequality holds if and only if γ is an ellipse.

4.2 The AGCSF for σ in $(1/3, 1)$

In this section, we generalize the Gage-Hamilton Theorem to $(4.1)_\sigma$ for $\sigma \in (1/3, 1)$. As we have seen in the previous chapter, the monotonicity of the entropy is essential in controlling the curvature of the normalized flow. Fortunately, a definition of entropy is available and monotonicity along the flow is still valid.

We define the **entropy** for the normalized flow (4.2) to be

$$\mathcal{E}\big(\tilde{\gamma}(\cdot, \tau)\big) = \left(\fint_{S^1} \Phi(\theta) \tilde{k}^{\sigma-1} d\theta \right)^{\frac{1}{\sigma-1}} , \quad \sigma \neq 1 \ .$$

When $\sigma = 1$, the entropy was defined in (3.17).

Lemma 4.6 *For all $\sigma > 0$,*

$$\frac{d}{dt} \mathcal{E}\big(\tilde{\gamma}(\cdot, \tau)\big) \leqslant 0 \ ,$$

and the equality holds if and only if $\gamma(\cdot, \tau)$ is a self-similar solution of (4.1) which contracts to some (x_0, y_0).

Proof: By (4.3),

$$\frac{d}{d\tau}\mathcal{E}\big(\widetilde{\gamma}(\cdot,\tau)\big)$$

$$= \frac{1}{2\pi}\Big(\oint \Phi(\theta)\widetilde{k}^{\sigma-1}d\theta\Big)^{\frac{2-\sigma}{\sigma-1}}\Big[\int \Phi\widetilde{k}^{\sigma}A[\Phi\widetilde{k}^{\sigma}]d\theta - \frac{1}{2A}\Big(\int \Phi\widetilde{k}^{\sigma}A[\widetilde{h}]d\theta\Big)^{2}\Big]$$

Noticing that

$$A = \frac{1}{2}\int \widetilde{h}A[\widetilde{h}]d\theta$$

is equal to π, it follows immediately from the Minkowski inequality that $d\mathcal{E}/dt(\widetilde{\gamma}(\cdot,\tau)) \leqslant 0$. Moreover, in case equality holds at some τ, then $\widetilde{\gamma}(\cdot,\tau)$ satisfies (4.4), where h is defined with respect to some (x_0, y_0). $\qquad\square$

Lemma 4.7 *For any σ in $(1/3,1)$, the isoperimetric ratio and the diameter of $\widetilde{\gamma}(\cdot,\tau)$ are uniformly bounded by a constant depending on its entropy.*

Proof: It suffices to show that the width $w(\theta)$ of $\widetilde{\gamma}(\cdot,\tau)$ along any direction θ has a uniform, positive lower bound. Consider $\sigma \geqslant 2/3$ first. By the Hölder inequality,

$$\int \widetilde{k}^{\sigma-1}d\alpha \leqslant \Big(\int \widetilde{k}^{-1}\big|\cos(\alpha - \theta)\big|d\alpha\Big)^{1-\sigma}\Big(\int \big|\cos(\alpha-\theta)\big|^{1-\frac{1}{\sigma}}d\alpha\Big)^{\sigma}.$$

Therefore,

$$w(\theta) \geqslant C_0\mathcal{E}_\sigma(\widetilde{\gamma}(\cdot,\tau))^{-1}.$$

Next, for $\sigma \in (1/3,2/3)$, by the Hölder inequality again,

$$\mathcal{E}_\sigma(\widetilde{\gamma}(\cdot,\tau)) \geqslant \mathcal{E}_{\frac{2}{3}}(\widetilde{\gamma}(\cdot,\tau))^{\frac{\sigma-\frac{1}{3}}{1-\sigma}}\mathcal{E}_{\frac{1}{3}}(\widetilde{\gamma}(\cdot,\tau))^{\frac{\frac{2}{3}-\sigma}{1-\sigma}}.$$

By the affine isoperimetric inequality and the definition of $\mathcal{E}_{\frac{1}{3}}$, we have

$$\mathcal{E}_{\frac{1}{3}}(\widetilde{\gamma}(\cdot,\tau)) \geqslant (2\pi\Phi_{\min})^{-3/2}.$$

Therefore,

$$\mathcal{E}_\sigma(\tilde{\gamma}(\cdot, 0)) \;\geqslant\; (2\pi\Phi_{\min})^{-3/2}\mathcal{E}_{\frac{2}{3}}\left(\tilde{\gamma}(\cdot, \tau)\right)^{\frac{\sigma-\frac{1}{3}}{1-\sigma}}$$

$$\geqslant\; C_0(2\pi\Phi_{\min})^{-3/2}w(\theta)^{\frac{\frac{1}{3}-\sigma}{1-\sigma}} \; .$$

$\square$

Lemma 4.8 *For $\sigma \in (1/3, 1)$, the curvature of $\tilde{\gamma}(\cdot, \tau)$ is uniformly pinched between two positive constants.*

Proof: We first derive an upper estimate. Consider the auxiliary function $\Phi = (-h_t)/(h - c(t) \cdot x - \delta)$ as in the proof of Lemma 3.15. Recall that $c(t)$ is the support center of h and $\delta = \rho r_0/2$. By following the same proof, we arrive at the estimate

$$k^\sigma \leqslant C(h - c(t) \cdot x - \delta) \; .$$

However, since now the isoperimetric ratio of $\gamma(\cdot, t)$ is uniformly bounded by Lemma 4.7, there exists a constant C' such that $h - c(t) \cdot x \leq C'r_{in}(t)$. Consequently, $\tilde{k}^\sigma$ is bounded from above.

Next, by the gradient estimate in Lemma 3.7, which holds for (4.3) after replacing $\tilde{k}$ by $\Phi\tilde{k}^\sigma$ and is scaling invariant, we know that

$$\left|(\tilde{k}^{\sigma-1})_{\tilde{s}}\right| = \left|\frac{1-\sigma}{\sigma}\right|(\tilde{k}^\sigma)_\theta \; ,$$

where $\tilde{s}$ is the arc-length of $\tilde{\gamma}$. Since the length of $\tilde{\gamma}(\cdot, \tau)$ is uniformly bounded and $\sigma < 1$, by integrating this inequality we obtain a positive lower bound for $\tilde{k}$. $\square$

Theorem 4.9 *For $\sigma \in (1/3, 1)$, any sequence $\{\tilde{\gamma}(\cdot, \tau_j)\}$, $\tau_j \longrightarrow \infty$, contains a subsequence which converges smoothly to a contracting self-similar solution of $(4.1)_\sigma$.*

Proof: We have obtained two-sided bounds on the normalized curvature in the preceding lemmas. By representing the equation for $\tilde{k}$ as polar graphs we can obtain all bounds on the derivatives of $\tilde{k}$ (see §3.4). Therefore, by Lemma 4.6,

$$\lim_{\tau \longrightarrow \infty} \frac{d\mathcal{E}}{d\tau}(\tilde{\gamma}(\cdot, \tau)) = 0$$

and any sequence $\{\tilde{\gamma}(\cdot, \tau)\}$ contains a subsequence converging smoothly to a convex curve γ_∞, which, according to the Minkowski inequality, satisfies

$$\Phi k_\infty^\sigma = \lambda h_\infty + \langle (x_0, y_0), (\cos\theta, \sin\theta) \rangle$$

for some positive λ and (x_0, y_0). If (x_0, y_0) is not the origin, from $A(t)^{1/2}\tilde{\gamma} = \pi^{1/2}\gamma$ we see that, for all t sufficiently close to ω, $\gamma(\cdot, t)$ no longer contain the origin. But this is impossible. Hence, (x_0, y_0) must be the origin. The proof of Theorem 4.9 is completed. □

Remark 4.10 When $\sigma > 1$, the isoperimetric ratio and the curvature of $\tilde{\gamma}(\cdot, \tau)$ are still bounded from above. This follows from combining the entropy inequality and the scaling invariant gradient estimate in Lemma 3.7 (replace k by Φk^σ). However, a positive lower bound for the curvature may not be available. Instead, one can derive a gradient estimate for the curvature directly using (4.3). In [10], the following theorem is proved.

Theorem *For $\sigma > 1$, any $\{\tilde{\gamma}(\cdot, \tau_j)\}, \tau_j \to \infty$, contains a subsequence which converges to a self-similar solution of (4.1) in $C^{k+2,\gamma}$-norm where $\gamma = (\sigma-1)^{-1} - k \in (0, 1]$. Moreover, the convergence is smooth away from points where the curvature of the self-similar solution is zero.*

An example in the same paper shows that the self-similar solution may not be strictly convex and the regularity concerning the subconvergence is optimal.

Remark 4.11 Where $\sigma < 1/3$, Theorem 4.9 holds under the additional assumption that the isoperimetric ratio of $\tilde{\gamma}(\cdot, \tau)$ is uniformly bounded. Again, this assumption cannot be fulfilled most times. To see why, we look at the equation for the self-similar solution (4.4) which is now written in the form of an elliptic equation,

$$h'' + h = \frac{a(\theta)}{h^p} \ , \tag{4.5}$$

where $a(\theta) = (\mu^{-1}\Phi)^{1/\sigma}$ and $p = 1/\sigma$. For simplicity, we shall consider the isotropic case $(a(\theta) \equiv 1)$ only. Self-similar solutions are critical points of the functional

$$\mathcal{F}(h) = \frac{1}{2} \int_{S^1} (h^2 - h_\theta^2) d\theta \int_{S^1} h^{1-p} d\theta \ .$$

Along the normalized flow, we have

$$\frac{d}{dt} \mathcal{F}(\tilde{h}(\cdot, \tau)) \ = \ (1-p) \int \tilde{h}^{-p} \tilde{h}_\tau d\theta$$

$$= \ (1-p) \Big(\int \tilde{k}^{\sigma-1} d\theta \int \tilde{h}^{-p+1} d\theta - \int \tilde{h}^{-p} \tilde{k}^\sigma d\theta \Big) \ .$$

By the Hölder inequality,

$$\int \tilde{k}^{\sigma-1} d\theta \leqslant \Big(\int \tilde{k}^\sigma h^{-p} d\theta \Big)^{\frac{\sigma}{\sigma+1}} \Big(\int \frac{h}{k} d\theta \Big)^{\frac{1}{\sigma+1}}$$

and

$$\int h^{\frac{\sigma-1}{\sigma}} d\theta \leqslant \Big(\int \tilde{k}^\sigma h^{-p} d\theta \Big)^{\frac{1}{\sigma+1}} \Big(\int \frac{h}{k} d\theta \Big)^{\frac{\sigma}{\sigma+1}} \ .$$

Therefore, for $\sigma \leqslant 1$,

$$\frac{d}{dt} \mathcal{F}(\tilde{h}(\cdot, \tau)) \geqslant 0 \ ,$$

and the equality holds if and only if $\tilde{h}$ is a self-similar solution.

When $\sigma \geqslant 1/3$, the Blaschke-Santaló's inequality asserts that the functional $\mathcal{I}$ has a universal upper bound. However, $\mathcal{F}$ becomes unbounded if $\sigma < 1/3$. To see this, let's consider the boundary of a rectangle (centered at the origin) whose dimension is 2ℓ by $\pi/2\ell$. Its support function is given by

$$h = \sqrt{\ell^2 + \frac{\pi^2}{\ell^2}} \begin{cases} \cos(\theta_0 - \theta) & , \ 0 \leqslant \theta \leqslant \theta_0 \\[2mm] \cos(\theta - \theta_0) & , \ \theta_0 \leqslant \theta \leqslant \pi/2 \ . \end{cases}$$

One can verify directly that, for $p > 3$,

$$\int_{S^1} h^{1-p} d\theta$$

tends to ∞ as $\ell \longrightarrow \infty$. On the other hand, one can show that, in the isotropic case, circles are the only self-similar solutions (or at least show that $\sup \mathcal{I}$ over all self-similar solutions is finite). Therefore, for any initial curve whose support function h_0 satisfies $\mathcal{F}(h_0) > \sup\{\mathcal{F}(h) : h \text{ is self-similar}\}$, the normalized flow starting at h_0 never subconverges to a self-similar solution. In fact, its isoperimetric ratio tends to infinity as $\tau \longrightarrow \infty$.

4.3   The affine curve shortening flow

We begin with a very brief description of the affine geometry of plane curves.

     Let γ be a plane curve. A parametrization of γ, σ, is called the affine arc-length parameter if it satisfies

$$[\gamma_\sigma, \gamma_{\sigma\sigma}] \equiv \det \begin{bmatrix} \gamma_\sigma^1 & \gamma_\sigma^2 \\[2mm] \gamma_{\sigma\sigma}^1 & \gamma_{\sigma\sigma}^2 \end{bmatrix}$$

$$= 1$$

along the curve. For any uniformly convex curve γ with a given parametrization p, its affine arc-length always exists and is given by

$$\sigma = \int_0^p [\gamma_p, \gamma_{pp}]^{1/3} dp .$$

Notice that, when γ is parametrized by the Euclidean arc-length s or the normal angle θ, the affine arc-length is given by

$$\sigma = \int_\gamma k^{1/3} ds$$

or

$$\sigma = \int_{S^1} k^{-2/3} d\theta .$$

The **affine arc-length of** γ (**affine perimeter** when γ is closed) is given by

$$\mathcal{L}(\gamma) = \int_\gamma d\sigma .$$

It has the invariance property, namely,

$$\mathcal{L}(\gamma_A) = \mathcal{L}(\gamma) ,$$

for any $\gamma_A \equiv A \cdot \gamma$ where $A \in SL(2, \mathbb{R})$ is any special affine transformation in the plane.

The **affine tangent** and **affine normal** of γ are given by $\mathcal{J} = \gamma_\sigma$ and $\mathcal{N} = \gamma_{\sigma\sigma}$, respectively. They satisfy $A\mathcal{J} = \mathcal{J}_A$ and $A\mathcal{N} = \mathcal{N}_A$ where $\mathcal{J}_A$ and $\mathcal{N}_A$ are, respectively, the affine tangent and normal for γ_A. By differentiating $[\gamma_\sigma, \gamma_{\sigma\sigma}] = 1$, we see that $\gamma_{\sigma\sigma\sigma}$ and γ_σ are linearly dependent and, hence, $\gamma_{\sigma\sigma\sigma} + \kappa\gamma_\sigma = 0$ for some κ. The function

$$\kappa = [\gamma_{\sigma\sigma} , \gamma_{\sigma\sigma\sigma}]$$

is the **affine curvature** of γ. It turns out that the affine curvature is an absolute invariant of the special affine group $SL(2, \mathbb{R})$ in the sense that the affine curvature of γ_A at $\gamma_A(\cdot)$ is equal to the affine curvature of γ at $\gamma(\cdot)$. Any function with this property is called an absolute invariant or a differential invariant. Other absolute invariants of $SL(2, \mathbb{R})$ include the derivatives of the affine curvature with respect to the affine arc-length, as one can check directly. In fact, any absolute invariant of $SL(2, \mathbb{R})$ must be a function of κ and its derivatives. Just as the Euclidean curvature determines the curve up to a Euclidean motion, the affine curvature also determines the curve up to a special affine transformation.

Example 4.12 Curves of constant affine curvature. As one can check directly, parabolas have zero affine curvature, and ellipses and hyperbolas have positive and negative affine curvature, respectively. Let $\gamma(\sigma) = (a \cos \lambda \sigma, \ b \sin \lambda \sigma)$ where $\lambda = (ab)^{1/3}$. Then, σ is the affine arc-length parameter and the affine curvature is given by $\kappa = (ab)^{-2/3}$. Similarly, let $C(\sigma) = (a \cosh(-\lambda)\sigma, \ b \sinh(-\lambda)\sigma)$ where $\lambda = -(ab)^{1/3}$. Then, σ is the affine arc-length parameter and $\kappa = -(ab)^{-2/3}$.

The area enclosed by the closed curve remains unchanged under all special affine transformations. Hence, the quotient

$$\frac{\mathcal{L}^3(\gamma)}{A} = \frac{\left(\int_{S^1} k^{-2/3} d\theta\right)^3}{A}$$

remains unchanged under all general affine transformations, that is, the full affine group $GL(2, \mathbb{R})$. The affine isoperimetric inequality asserts that only the ellipses maximize this quotient.

Now, let's consider the flow

$$\frac{d\gamma}{dt} = \mathcal{N} \ , \tag{4.6}$$

where $\mathcal{N}$ is the affine normal of $\gamma(\cdot, t)$. Since $\mathcal{N}_A = A\mathcal{N}$, we see that, if $\gamma(\cdot, t)$ solves (4.5), so does $\gamma_A(\cdot, t)$ for any $A \in SL(2, \mathbb{R})$. In particular, all ellipses are contracting self-similar solutions for (4.6).

Given (4.6), one may derive the corresponding evolution equations for the geometric quantities of the flow. Since the derivation is somehow parallel to the Euclidean case which was done in §1.3, we list the result only. One may consult [98] for details. We have

$$\sigma_t = -\frac{2}{3}\kappa\sigma , \tag{4.7}$$

$$\mathcal{T}_t = -\frac{1}{3}\kappa\mathcal{T} ,$$

$$\mathcal{N}_t = \frac{1}{3}\kappa\mathcal{N} - \frac{1}{3}\kappa_s\mathcal{T} ,$$

and

$$\kappa_t = \frac{1}{3}\kappa_{\sigma\sigma} + \frac{4}{3}\kappa^2 . \tag{4.8}$$

We also have

$$A_t = -\mathcal{L} \text{ and} \tag{4.9}$$

$$\mathcal{L}_t = -\frac{2}{3}\int_\gamma \kappa d\sigma . \tag{4.10}$$

Sometimes it is convenient to express σ and κ as functions of the normal angle θ. In fact, we can verify that

$$d\sigma = v^{-2}d\theta$$

and

$$\kappa = v^3(v_{\theta\theta} + v) ,$$

where $v \equiv k^{1/3}$.

Let's express the affine normal in terms of the Euclidean tangent and normal. We have

$$\mathcal{N} = k^{-1/3}\gamma_{\sigma s}$$

$$= k^{-1/3}(k^{-1/3}\gamma_s)_s$$

$$= k^{1/3}\boldsymbol{n} + k^{-1/3}(k^{-1/3})_s\boldsymbol{t} \ .$$

Hence, by Proposition 1.1, the flow (4.6) is equivalent to $(4.1)_\sigma$ with $\sigma = 1/3$ and $\Phi \equiv 1$.

Theorem 4.13 *The normalized flow $\widetilde{\gamma}(\cdot,\tau)$ of $(4.1)_\sigma$, $\sigma = 1/3$ and $\Phi \equiv 1$, converges smoothly to an ellipse centered at the origin.*

Lemma 4.14 *We have*

$$\frac{d}{d\tau}\mathcal{L}^3(\widetilde{\gamma}(\cdot,\tau)) \geqslant 0 \ ,$$

and equality if and only if $\widetilde{\gamma}(\cdot,\tau)$ is an ellipse.

Proof: This is nothing but the monotonicity property of the entropy in Lemma 4.6 for $\sigma = 1/3$. □

Let $\tau_1 \in (\tau_0, \infty)$ be fixed. We choose a special affine transformation A such that

$$\Gamma_0 = A\widetilde{\gamma}(\cdot,\tau_1)$$

is bounded between the circles $C_{R^{-1}}((0,0))$, and $C_R((0,0))$ where R is an absolute constant greater than 1. (In fact, one can take $R = 3$.) Observe that the circle $C_{R^{-1}}((0,0))$ shrinks to a point under the affine flow. Let T be the time it reaches $C_{1/(2R)}((0,0))$. By the strong separation principle, we know that $\Gamma(\cdot,t)$, the affine flow starting at Γ_0, is bounded between $C_{1/(2R)}((0,0))$ and $C_R((0,0))$ for all t in $[0, T]$. Letting H be the support function of Γ, we consider the auxiliary function

$$\Phi = \frac{tH(\theta, t)}{H(\theta, t) - \rho} \ , \quad t \in [0, T] \ ,$$

where $\rho = 1/(4R)$. Arguing as in the proof of Proposition 3.3, we obtain the inequality

$$\frac{\rho K^{5/3} t_0}{3(H - \rho)} \leqslant \frac{4}{3} \frac{K^{2/3} t_0}{H - \rho} + K^{1/3} + K^{2/3} \ ,$$

where K (the curvature of Γ) and H are evaluated at (θ_0, t_0), a maximum of Φ. It follows that

$$K(\theta, t) \leqslant \frac{C}{t} \ , \quad \forall (\theta, t) \in S^1 \times (0, T] \ .$$

In particular, K has a uniform upper bound on $[T/2, T]$. For any large τ, we can find $\tau_1 < \tau$, τ_1 close to τ such that the flow $\Gamma(\cdot, t)$ starting at $A\gamma(\cdot, \tau_1)$ as described above satisfies $\tilde{\gamma}(\cdot, \tau) = A^{-1}\Gamma(\cdot, T')$ for some T' in $[T/2, T]$. Since $\Gamma(\cdot, T')$ is bounded between $C_{1/(4R)}\big((0, 0)\big)$ and $C_R\big((0, 0)\big)$, we can multiply it by a constant so that its enclosed area is π. We have shown that, for each large τ, there exists $A_\tau \in SL(2, \mathbb{R})$ such that $\overline{\gamma}(\cdot, \tau) = A_\tau \tilde{\gamma}(\cdot, \tau)$ has uniformly bounded curvature.

Next, we claim that $\overline{k}$, the curvature of $\overline{\gamma}$, also admits a positive lower bound when τ is sufficiently large. First, we note that, by applying the maximum principle to (4.8),

$$\kappa(\cdot, t) \geqslant \inf \kappa(\cdot, 0) \ ,$$

i.e.,

$$\tilde{\kappa}(\cdot, \tau) \geqslant \left(\frac{A(t)}{\pi}\right)^{\frac{2}{3}} \inf \kappa(\cdot, 0) \ .$$

Using the affine invariance of the affine curvature, the affine curvature of $\overline{\gamma}$, satisfies

$$\overline{\kappa}(\cdot, \tau) \geqslant -C_0 A(t)^{2/3} \ .$$

We rewrite this inequality as

$$A[V + C_0 A(t)^{2/3} H] \geqslant 0 ,$$

where we have set $V = \bar{k}^{1/3}$. Hence, $\overline{H} \equiv V + C_0 A^{2/3}(t)H$ defines a convex curve $\overline{\Gamma}$. By the formula

$$L(\overline{\Gamma}) = \int_{S^1} \frac{1}{\overline{K}(\theta,\tau)} d\theta ,$$

we deduce that for any $\sigma > 0$,

$$\left|\{\overline{K} \leqslant \sigma\}\right| \;\leqslant\; \sigma L(\overline{\Gamma})$$

$$\leqslant\; \sigma \times 2\pi R .$$

Notice that, by convexity, the perimeter of $\overline{\Gamma}$ is less than the perimeter of $C_R\big((0,0)\big)$. It follows from the definition of $\overline{H}$ that

$$\left|\{\overline{H} \leqslant \sigma\}\right| \leqslant \sigma^3 (2\pi R)^3 . \tag{4.11}$$

On the other hand, let $\overline{\Gamma}(\theta_0,\tau)$ be a point in the set $\{\overline{H} < \sigma\}$. We have, for all θ satisfying $\cos(\theta - \theta_0) \geqslant 0$,

$$\overline{H}(\theta) \leqslant \sigma \cos(\theta - \theta_0) + d\sqrt{1 - \cos(\theta - \theta_0)^2} .$$

This is because the right-hand side of this inequality is nothing but the support function of the region lying inside a cylinder of width d and with axis along $(\cos\theta_0, \sin\theta_0)$, truncated by the line $\{\cos\theta_0 x + \sin\theta_0 y = \sigma\}$. When d is larger than the maximal width of $\overline{\Gamma}$, this region contains $\overline{\Gamma}$ and, hence, the inequality holds. It follows that

$$\left\{\theta : \left|\sin(\theta - \theta_0)\right| \leqslant \sigma\right\} \subseteq \left\{\overline{H} \leqslant (1 + d)\sigma\right\} ,$$

and so

$$\left|\{\overline{H} \leqslant (1 + d)\sigma\}\right| \geqslant 2\sigma . \tag{4.12}$$

By putting (4.10) and (4.11) together, we see that the set $\{\overline{H} \leqslant \sigma\}$ is empty if $\sigma < 2^{1/2}(1+d)^{-3/2}(2\pi R)^{-3/2}$. From

$$\overline{k}^{1/3} = V \geqslant -C_0 A(t)^{2/3} H + 2\pi^{-3/2} R^{-3/2} \ ,$$

we deduce a positive lower bound for $\overline{k}$ when t is sufficiently close to ω. We remark that such a bound can also be derived from the upper bound on K by using the Harnack's inequality of Chow [37].

With two-sided bounds on K, we can use parabolic regularity theory to obtain bounds on the derivatives of K (in θ) for all sufficiently large τ_1. We have proved the following lemma.

Lemma 4.15 *For each τ, there exists $A_\tau \in SL(2, \mathbb{R})$ so that $\overline{\gamma}(\cdot, \tau) = A_\tau \widetilde{\gamma}(\cdot, \tau)$ satisfies (a) its curvature is uniformly pinched between two positive constants and (b) the derivatives of its curvature are uniformly bounded.*

We note that three useful messages are encoded in this lemma. First, by affine invariance, the affine curvature of $\widetilde{\gamma}$, as well as its derivatives with respect to the affine arc-length, is uniformly bounded. Next, since by Lemma 4.14 we know that there exists a sequence $\{\overline{\gamma}(\cdot, \tau_j)\}$, $\tau_j \longrightarrow \infty$, such that $\overline{\kappa}(\cdot, \tau_j)$ tend to 1, $\widetilde{\kappa}(\cdot, \tau_j)$ also tends to 1. Further, by applying the maximum principle to (4.8), we conclude that

$$\kappa(\theta, t) \geqslant \inf_\theta \kappa(\theta, t_j) \ , \quad t > t_j \ ,$$

and so, $\kappa(\cdot, t)$ is positive for all t close to ω.

Lemma 4.16 $\lim_{\tau \longrightarrow \infty} \widetilde{\kappa}(\cdot, \tau) = 1$.

Proof: Consider the fully affine invariant quantity

$$\mathcal{L}(t) \int \kappa d\sigma \ .$$

Notice that, by the Minkowski inequality,

$$\int \kappa d\sigma \;\leqslant\; \frac{1}{2}\,\mathcal{L}^2/A\;,$$

$$\mathcal{L}\int \kappa d\sigma \;\leqslant\; \frac{1}{2}\frac{\mathcal{L}^3}{A}$$

$$\leqslant\; 4\pi^2\;,$$

and equality holds if and only if the curve is an ellipse. By (4.7), (4.8), (4.10), and the Hölder inequality,

$$\frac{d}{dt}\Big(\mathcal{L}(t)\int \kappa d\sigma\Big) \;=\; \frac{2}{3}\Big[-\Big(\int \kappa d\sigma\Big)^2 + \mathcal{L}\int \kappa^2 d\sigma\Big]$$

$$\geqslant\; 0\;.$$

Hence in the equality

$$A\int \kappa d\sigma - \frac{1}{2}\mathcal{L}^2 = \frac{A}{\mathcal{L}}\Big(\mathcal{L}\int \kappa d\sigma - \frac{\mathcal{L}^3}{2A}\Big)\;,$$

both terms $\mathcal{L}\int \kappa d\sigma$ and $\mathcal{L}^3/(2A)$ increase to $4\pi^2$ as $t \uparrow \omega$. We conclude

$$\frac{1}{2}\tilde{\mathcal{L}}^2 - \pi\int \tilde{\kappa}d\tilde{\sigma} \to 0 \quad \text{as } \tau \to \infty\;.$$

By affine invariance, we also have

$$\frac{1}{2}\overline{\mathcal{L}}^2 - \pi\int \overline{\kappa}d\overline{\sigma} = o(1)$$

as $\tau \to \infty$. Now, let's apply Theorem 4.3 to the functions $h = \overline{h}$ and $\varphi = \overline{k}^{1/3}$ to get

$$\lim_{\tau\to\infty} \int |\overline{h} - \overline{k}^{1/3}|^2 d\theta = 0\;.$$

In view of Lemma 4.15, we deduce

$$\lim_{\tau\to\infty} \overline{h}\,\overline{k}^{-1/3} = 1\;. \tag{4.13}$$

Now, we convert the information on $\bar{\gamma}$ to an estimate on $\tilde{\kappa}$ as follows. One can verify directly that

$$\left(\bar{h}\,\bar{k}^{-1/3}\right)_{\bar{\sigma}\,\bar{\sigma}} + \bar{\kappa}(\bar{h}\,\bar{k}^{-1/3}) = 1 \ .$$

By (4.13) and interpolation,

$$\lim_{\tau \to \infty} (\bar{h}\,\bar{k}^{-1/3})_{\bar{\sigma}\,\bar{\sigma}} = 0 \ .$$

Therefore, $\lim\limits_{\tau \to \infty} \tilde{\kappa} = \lim\limits_{\tau \to \infty} \bar{\kappa} = 1$. The proof of Lemma 4.16 is completed. $\square$

Now we can prove Theorem 4.13. By a direct computation, the Euclidean curvature and the affine curvature of the normalized affine CSF satisfy

$$\tilde{k}_\tau = \left(\tilde{\kappa} - \frac{\tilde{\mathcal{L}}}{2\pi}\right)\tilde{k} \tag{4.14}$$

and

$$\tilde{\kappa}_\tau = \frac{1}{3}\tilde{\kappa}_{\tilde{\sigma}\tilde{\sigma}} + \frac{4}{3}\left(\tilde{\kappa} - \frac{\tilde{\mathcal{L}}}{2\pi}\right)\tilde{\kappa} \ , \tag{4.15}$$

respectively. Since $\tilde{\kappa}$ and $\tilde{\kappa} - \tilde{\mathcal{L}}/2\pi$ tend to 1 and 0, respectively, as $\tau \to \infty$, for each large τ', $\tilde{\gamma}(\cdot, \tau)$ is very close to an ellipse in some time interval (τ', τ''). By applying a special affine transformation to make $\tilde{\gamma}(\cdot, \tau')$ almost circular, we may assume that $\tilde{\sigma}$, its affine arclength, is very close to 1 in (τ', τ''). Here $\tau'' - \tau'$ can be arbitrarily large as τ' tends to ∞. Moreover, from the formula

$$\tilde{v}^3\left(\tilde{v}_{\theta\theta} + \tilde{v}\right) = \tilde{\kappa} \ , \quad \tilde{v} = \tilde{\sigma}^{-1/2} \ ,$$

we see that $\tilde{\sigma}_\theta$ and $\tilde{\sigma}_{\theta\theta}$ are very small in this interval. For a small ε_0 to be chosen later, let (τ', τ'') be the largest open interval on which

$$1 - \varepsilon_0 \leqslant \sigma \leqslant 1 + \varepsilon_0$$

and

$$|\tilde{\sigma}_\theta, \tilde{\sigma}_{\theta\theta}| \leqslant \varepsilon_0 . \tag{4.16}$$

We may compare (4.15) with the heat equation

$$u_\tau = \frac{1}{3} u_{\theta\theta} ,$$

which satisfies $osc\; u = O(exp(-\tau/3))$. For some fixed small ε_0,

$$osc\; \tilde{\kappa} \leqslant Ce^{-\frac{1}{6}\tau} , \;\; \forall \tau \in (\tau', \tau'') .$$

Substituting this estimate into (4.14) and then integrating the inequality, we have, for $\tau_2 > \tau_1$ in $[\tau', \tau'')$,

$$\left| \log \frac{\tilde{k}(\theta, \tau_2)}{\tilde{k}(\theta, \tau_1)} \right| \leqslant C'e^{-\frac{1}{6}\tau_1} . \tag{4.17}$$

Taking $\tau_1 = \tau'$, we conclude that $\tilde{k}(\theta, \tau_2)$ is arbitrarily close to 1. Hence, (4.16) holds for all $\tau \geqslant \tau'$ and so $\tau'' = \infty$. Now (4.17) implies that $\tilde{\gamma}(\cdot, \tau)$ converges uniformly to some γ_∞ as $\tau \to \infty$. By Lemma 4.16, γ_∞ must be an ellipse. Finally, following the end of the proof in Theorem 4.9, we know that γ_∞ is centered at the origin. The proof of Theorem 4.13 is completed.

4.4   Uniqueness of self-similar solutions

In this section, we shall prove the following uniqueness result.

Theorem 4.17 *Under the assumptions $\sigma \geqslant 1$ and $\Phi(\theta + \pi) = \Phi(\theta)$, the equation $\Phi k^\sigma = h$ has a unique solution which satisfies $h(\theta + \pi) = h(\theta)$.*

Proof: To show that the solution exists, we solve (4.1)$_\sigma$ with a centrally symmetric γ_0. By Theorem 3.12(i) (for $\sigma = 1$) and the theorem in Remark 4.10, we know that its normalized flow subconverges to a convex curve which satisfies (4.4). After a suitable scaling, we can take $\mu = 1$. The uniqueness of this equation is contained in the following theorem. $\square$

Theorem 4.18 *Let φ be a contracting self-similar solution of the flow $(4.1)_\sigma$ satisfying (4.4) with $\mu = 1$, $\sigma \geqslant 1$, and symmetric Φ. Then, the relative isoperimetric ratio*

$$\frac{\left(\int hA[\varphi]d\theta\right)^2}{\frac{1}{2}\int hA[h]d\theta} \equiv \frac{L_\varphi^2}{A}$$

is strictly decreasing along any solution $\gamma(\cdot, t)$ of $(4.1)_\sigma$ (here h is the support function of $\gamma(\cdot, t)$), unless $h = \lambda\varphi + \langle(x_0, y_0), (\cos\theta, \sin\theta)\rangle$ for some positive λ and (x_0, y_0).

Proof: We have

$$\frac{d}{dt}\left(\frac{L_\varphi^2}{A}\right)$$

$$= \frac{d}{dt}\left(\frac{A_{10}^2}{A_0}\right)$$

$$= -2\frac{A_{10}}{A_0^2}\left[A_0\int \varphi A[\varphi]\left(\frac{A[\varphi]}{A[h]}\right)^\sigma d\theta - A_{10}\int \varphi A[\varphi]\left(\frac{A[\varphi]}{A(h)}\right)^{\sigma-1} d\theta\right].$$

By the Hölder inequality,

$$\frac{\int \varphi A[\varphi]\left(\frac{A[\varphi]}{A[h]}\right)^\sigma d\theta}{\int \varphi A[\varphi]\left(\frac{A[\varphi]}{A[h]}\right)^{\sigma-1} d\theta} \geqslant \frac{\int \varphi A[\varphi]\left(\frac{A[\varphi]}{A[h]}\right) d\theta}{A_1},$$

and so it suffices to prove the theorem for $\sigma = 1$:

$$\int \varphi A[\varphi] \left(\frac{A[\varphi]}{A[h]} \right) d\theta \geq \frac{A_1 A_{10}}{A_0} . \tag{4.18}$$

By the Hölder inequality again,

$$\int \varphi A[\varphi] \frac{A[\varphi]}{A[h]} d\theta \geq \frac{\left(\int h A[\varphi] d\theta \right)^2}{\int \frac{h^2 A[h]}{\varphi} d\theta} .$$

So (4.14) follows from

$$A_{10} A_0 \geq A_1 \int \frac{h^2}{\varphi} A[h] d\theta .$$

Let's assume for this moment h is symmetric, i.e., $h(\theta + \pi, t) = h(\theta, t)$. Then we can take $\rho = h/\varphi$ in the anisotropic Bonnesen inequality Theorem 4.2,

$$\left(\frac{h}{\varphi} \right)^2 A_1 - 2 \left(\frac{h}{\varphi} \right) A_{01} + A_0 \leq 0 .$$

Multiplying this inequality by $\varphi A[h]$ and then integrating over S^1, we have

$$A_1 \int \frac{h^2}{\varphi} A[h] d\theta - 2 A_{01} A_0 + A_0 A_{10} \leq 0 ,$$

which is precisely the desired estimate. Furthermore, we know that equality holds if and only if $h = \lambda \varphi + \langle (x_0, y_0), (\cos \theta, \sin \theta) \rangle$ for some positive λ and (x_0, y_0).

To prove (4.14) for non-symmetric curves, we proceed as follows. Mark a point $X(\theta)$ on the curve γ and let $Y(\theta)$ be the point on γ whose outer normal is $(\cos(\theta + \pi), \sin(\theta + \pi))$. The line segment connecting X and Y divides the region enclosed by γ into two subregions D_1 and D_2. As $X(\theta)$ transverses along the curve γ until the positions of $X(\theta)$ and $Y(\theta)$ are exchanged, the subregions change from D_1 and

D_2 to D_2 and D_1. By the mean-value theorem, there exists some $X(\theta_0)$ such that $|D_1| = |D_2|$. The line segment $\overline{X(\theta_0)Y(\theta_0)}$ divides γ into two arcs γ_+ and γ_-. We can extend γ_1 and γ_2 separately to get two closed convex curves $\overline{\gamma}_+$ and $\overline{\gamma}_-$ which are centrally symmetric with respect to the mid-point of $\overline{X(\theta_0)Y(\theta_0)}$. Now (4.14) holds for $\overline{\gamma}_+$ and $\overline{\gamma}_-$. (Notice that $\overline{\gamma}_+$ and $\overline{\gamma}_-$ are C^1-curves in general. However, Theorem 4.2 continues to hold.) Therefore, in obvious notation,

$$
\frac{\displaystyle\int \varphi A[\varphi]\frac{A[\varphi]}{A[h]}\,d\theta}{A_1} - \frac{A_{10}}{A_0}
$$

$$
= \frac{1}{2}\left(\frac{\displaystyle\int \varphi A[\varphi]\frac{A[\varphi]}{A[h_+]}\,d\theta}{A_1} - \frac{A_{10}^+}{A_0^+}\right) + \frac{1}{2}\left(\frac{\displaystyle\int \varphi A[\varphi]\frac{A[\varphi]}{A[h_-]}\,d\theta}{A_1} - \frac{A_{10}^-}{A_0^-}\right)
$$

$$
\leqslant 0 .
$$

$\square$

Notes

The results in Sections 1,2 and 4 are largely taken from Andrews [10], to which we refer for a systematic and rather complete discussion on the AGCSF. Proofs of the results on convex bodies stated in Section 1 can be found in Schneider [99]. Specifically, see §6.6 for the Minkowski inequality, (6.6.10) for Theorem 4.3, and (6.2.21) for the Bonnesen-type inequality. The Blaschke-Santaló inequality, which was due to Blaschke in the planar case, was proved in Blaschke [21]. (See also [2].) In the proof of subconvergence, one frequently uses the monotonicity of some functionals such as the entropy and the "Firey functional" $\mathcal{F}$. A nice discussion on these quantities can be found in Andrews [9]. The monotonicity of the relative isoperimetric ratio along the flow, Theorem 4.18, was first proved by Gage [54] for the CSF. Uniqueness of the self-similar solution for the symmetric, anisotropic CSF was first pointed out in Gage [57], where it is deduced from a classical theorem of Wulff. Another proof can be found in Dohmen-Giga [44]. Theorems 4.17 and 4.18 are taken from [10].

Expanding Flows. Most flows studied in this book are shrinking. We point out expanding isotropic and anisotropic flows for curves have been studied by several authors, see Urbas [108] [109], Gerhardt [60], Chow-Tsai [38], and Andrews [10].

Self-similar solutions for the AGCSF. One consequence of the subconvergence of the normalized flow of $(4.1)_\sigma$ is the existence of contracting self-similar solutions. According to Theorem 4.9 and Remark 4.10, they exist for $\sigma \in (1/3, \infty)$. Moreover, they are unique when Φ is symmetric and $\sigma \geqslant 1$. This issue can also be studied by looking at the elliptic equation (4.5) directly. In fact, using the degree-theoretic method, Dohmen, Giga, and Mizoguchi [45] proved the existence of these solutions for $\sigma \geqslant 1/2$. In Ai-Chou-Wei [2], we use the variational method to prove existence for $\sigma > 1/3$. Moreover, similarity is drawn between the Nirenberg problem and the equation $(4.5)_\sigma$ when $\sigma = 1/3$. By observing that the special affine group $SL(2, \mathbb{R})$ acts on S^1 like the conformal group acts on S^2 in the Nirenberg problem (Chang-Gursky-Yang [26]), necessary conditions and sufficient conditions for the existence of self-similar solutions were found. For instance, a necessary condition is the "Kazdan-Warner condition",

$$\int_{S^1} \frac{a'(\theta)e^{i\theta}}{h^2(\theta)} d\theta = 0 .$$

In particular, when $a = 1 + \varepsilon \cos \theta, |\varepsilon| < 1$, $(4.1)_{1/3}$ does not have any self-similar solution. On the other hand, we have:

Theorem. *Let*

$$a(\theta) = a_0 + \sum_{n=1}^{N} (a_n \cos 2n\theta + b_n \sin 2n\theta) > 0 .$$

Suppose that

$$\sum_{n=1}^{N-1} n|c_n| < N|c_N|, \qquad c_n = a_n + ib_n.$$

Then (4.5) has at least $(N-1)$ many solutions.

When $\sigma \in (0,1)$, it is pointed out in [10] that self-similar solutions are, in general, not unique. In Chou-Zhang [33], we show that stable self-similar solutions are unique for a class of anisotropic factors.

The affine curve shortening problem. Our discussion on the affine CSF basically follows Andrews [8], where a hypersurface driven by its affine normal is studied. The problem was also studied in Sapiro-Tannenbaum [98]. From a completely different context, the flow arises from image processing and computer vision, where people search for geometric flows which preserve ellipses. In Alvarez-Guichard-Lions-Morel [6], the flow, termed as the fundamental equations in image processing, was derived by the axiomatic method. On the other hand, it was proposed by Sapiro-Tannenbaum [97] in the context of affine geometry. A general result in [94] characterizes the affine CSF as the affine invariant curvature flow of the lowest order.

Curve flows in Klein geometry. As a further interesting development of the affine CSF, Olver-Sapiro-Tannenbum [94] propose the study of invariant curve flows in a Klein geometry (Guggenheimer [70] and Olver [93]). According to the Erlangen Programme, each Lie transformation group G acting on the plane gives rise to a geometry. For a given Lie group or a Lie algebra $\mathfrak{g}$ (realized as vector fields on $\mathbb{R}^2$), we can define its *group length element* $d\sigma$ to be the $\mathfrak{g}$-invariant one-form of lowest order, and the *group curvature* κ to be the absolute differential invariant of lowest order. It is known that all differential invariants of $\mathfrak{g}$ are functions of κ and its derivative with respect to the group length. Next, one define, the *group tangent* and *group normal* of a curve to be $\mathcal{J} = \gamma_\sigma$ and $\mathcal{N} = \gamma_{\sigma\sigma}$, respectively. Now the "group curve shortening flow" is

$$\frac{\partial\gamma}{\partial t} = \mathcal{N} \ . \tag{4.19}$$

Notice that, when the group is the Euclidean group\special affine group, the flow is the CSF\affine CSF. So (4.19) is a generalization of the curve shortening flow to other geometries. The expression for the group curvature for the similarity group, the fully affine group, and the projection group can be found in [94]. The reader should be aware that, for some groups, such as the similarity group, the flow is not curve shortening and also sometimes one needs to modify the equation to retain parabolicity.

As one can see from the variation formula (4.10) for the affine length, the ACSF is not pointwise shortening (it is affine length shortening, though [98]). To get a more natural "curve shortening flow" from the variational point of view, one should look at the L^2-gradient flow,

$$\frac{\partial\gamma}{\partial t} = -\kappa\mathcal{N} \ , \tag{4.20}$$

where we have put a minus sign before the affine curvature to guarantee parabolicity. This is a fourth-order parabolic equation. The affine length increases pointwisely along this flow. Notice that the isoperimetric ratio attains its minima at circles but the affine isoperimetric ratio attains its maxima at ellipses. It is proved in Andrews [11] that this curve lengthening flow expands to infinity and tends to an ellipse after suitable normalization.

Chapter 5

The Non-convex Curve Shortening Flow

In this chapter we first prove the Grayson convexity theorem.

Theorem 5.1 (Grayson convexity theorem) *Consider the CSF where γ_0 is a smooth, embedded closed curve. There exists a $t_0 < \omega$ such that $\gamma(\cdot, t)$ is uniformly convex for all $t \in [t_0, \omega)$.*

The proof of this theorem is based on the monotoncity of an isoperimetric ratio introduced by Hamilton (Section 1) and a blow-up argument (Section 2). Next, we discuss the classification of the singularities of the CSF for immersed curves in Section 3.

5.1 An isoperimetric ratio

Let γ be a smooth, embedded closed curve and D the region it encloses. We can always find some line segment Γ in $\overline{D}$ whose endpoints touch γ. It divides D into two subsets D_1 and D_2. When Γ does not touch γ away from its endpoints, both D_1 and D_2 are regions. In any case, we can associate to each triple (Γ, D_1, D_2) a number

$$G(\Gamma) \;=\; L^2\Big(\frac{1}{A_1} + \frac{1}{A_2}\Big)\,,$$

where L is the length of Γ and A_i is the area of $D_i(i = 1, 2)$. We define $g = g(\gamma)$ to be the infimum of $G(\Gamma)$ over all these admissible triples.

It is not hard to see that $g(\gamma)$ is always attained. Let (Γ_0, D_1, D_2) be a minimum of $g(\gamma)$. We first assert that $\Gamma_0 \cap \gamma$ only consists of two components containing the two endpoints separately. For, if there is a point in $\Gamma_0 \cap \gamma$ lying outside these two components, it divides Γ into two line segments Γ_1 and Γ_2. We consider the two admissible triples $(\Gamma_1, D_{11}, D_2 \bigcup D_{12})$ and $(\Gamma_2, D_{12}, D_2 \bigcup D_{11})$, where $D_{11} \bigcup D_{12} = D_1$ and $\Gamma_i \subseteq \partial D_{1i}$ $(i = 1, 2)$. We have

$$g(\gamma) = G(\Gamma_0) \leqslant G(\Gamma_i) , \quad i = 1, 2 .$$

From the definition of G we have, in obvious notation,

$$\frac{A_{11}(A_2 + A_{12})}{A_2(A_{11} + A_{12})} \leqslant \frac{L_1^2}{(L_1 + L_2)^2}$$

and

$$\frac{A_{12}(A_2 + A_{11})}{A_2(A_{11} + A_{12})} \leqslant \frac{L_2^2}{(L_1 + L_2)^2} .$$

Adding up these inequalities yields a contradiction:

$$1 + \frac{2A_{11}A_{12}}{A_2(A_{11} + A_{12})} \leqslant 1 - \frac{2L_1L_2}{(L_1 + L_2)^2} .$$

Hence, there are exactly two components in $\Gamma_0 \cap \gamma$.

Next, Γ_0 must be transversal to γ at the endpoints. For, if not, we can easily find a variation of Γ_0 in which the length decreases at an infinite rate while the area changes at a finite rate. In summary, any minimum Γ_0 of $g(\gamma)$ must intersect γ transversally, and $\Gamma_0 \cap \gamma$ consists of the two endpoints of Γ_0.

Now we compute the first and second variations at a minimum Γ_0. Without loss of generality, we may assume Γ_0 is a vertical line segment over $x = x_0$. Near its top (resp. its bottom) γ is represented as the graph of a function $y = y^+(x)$ (resp. $y = y^-(x)$). For any fixed real numbers a and b, consider the line segment Γ_μ which connects $(x_0 + a\mu, y^-(x_0 + a\mu))$ and $(x_0 + b\mu, y^+(x_0 + b\mu))$ for small μ. The square of the length of Γ_μ and the area of the region lying on the left of Γ_μ are given by

$$L^2(\mu) = (b-a)^2\mu^2 + \left(y^+(x_0 + b\mu) - y^-(x_0 + a\mu)\right)^2$$

and

$$A_1(\mu) = A_1(0) + \int_{x_0}^{x_0+b\mu} y^+(x)dx - \int_{x_0}^{x_0+a\mu} y^-(x)dx$$
$$- \frac{1}{2}\left(y^+(x_0 + b\mu) + y^-(x_0 + a\mu)\right)(b-a)\mu ,$$

respectively. We have

$$\frac{dL}{d\mu} = by_x^+(x_0) - ay_x^-(x_0) ,$$

$$\frac{d^2L}{d\mu^2} = \frac{1}{L}\left[(b-a)^2 + L\left(b^2 y_{xx}^+(x_0) - a^2 y_{xx}^-(x_0)\right)\right] ,$$

$$\frac{dA_1}{d\mu} = \frac{b+a}{2}\left(y^+(x_0) - y^-(x_0)\right) ,$$

and

$$\frac{d^2A_1}{d\mu^2} = ab\left(y_x^+(x_0) - y_x^-(x_0)\right) ,$$

at $\mu = 0$. Now, as the function $G(\Gamma_\mu)$ attains minimum at $\mu = 0$, we have

$$0 = \frac{d}{d\mu}\Big|_{\mu=0} \log L^2(\mu) \Big(\frac{1}{A_1(\mu)} + \frac{1}{A - A_1(\mu)}\Big)$$

$$= \frac{2}{L}\big(by_x^+(x_0) - ay_x^-(x_0)\big) - \frac{b+a}{2}\Big(\frac{1}{A_1} - \frac{1}{A - A_1}\Big)\big(y^+(x_0) - y^-(x_0)\big)$$

Taking $a = -b = 1$ we deduce

$$y_x^+(x_0) = -y_x^-(x_0) \ . \tag{5.1}$$

Putting this back to the above expression, we have

$$y_x^+(x_0) = \frac{L^2}{4}\Big(\frac{1}{A_1} - \frac{1}{A - A_1}\Big) \ . \tag{5.2}$$

Next we compute

$$0 \leqslant \frac{d^2}{d\mu^2}\Big|_{\mu=0} \log L^2(\mu)\Big(\frac{1}{A_1(\mu)} + \frac{1}{A - A_1(\mu)}\Big)$$

$$= \frac{2}{L^2}\Big[(b-a)^2 + L\big(b^2 y_{xx}^+(x_0) - a^2 y_{xx}^-(x_0)\big)\Big]$$

$$- \frac{2}{L^2}\big(by_x^+(x_0) - ay_x^-(x_0)\big)^2$$

$$- ab\Big(\frac{1}{A_1} - \frac{1}{A - A_1}\Big)\big(y_x^+(x_0) - y_x^-(x_0)\big)$$

$$+ \Big[\frac{1}{A_1^2} + \frac{1}{(A - A_1)^2}\Big]\Big(\frac{b+a}{2}\Big)^2\big(y^+(x_0) - y^-(x_0)\big)^2 \ .$$

Taking $a = b = 1$ and using (5.1) and (5.2) in this expression, we have

$$
0 \;\leqslant\; \frac{2}{L}\big(y_{xx}^+(x_0) - y_{xx}^-(x_0)\big) - \frac{2}{L^2}\big(y_x^+(x_0) - y_x^-(x_0)\big)^2
$$

$$
- \left(\frac{1}{A_1} - \frac{1}{(A - A_1)}\right)\big(y_x^+(x_0) - y_x^-(x_0)\big)
$$

$$
+ \left[\frac{1}{A_1^2} + \frac{1}{(A - A_1)^2}\right]\big(y^+(x_0) - y^-(x_0)\big)^2
$$

$$
\leqslant\; \frac{2}{L}\big(y_{xx}^+(x_0) - y_{xx}^-(x_0)\big) - L^2\left(\frac{1}{A_1} - \frac{1}{A - A_1}\right)^2
$$

$$
+ L^2\left[\frac{1}{A_1^2} + \frac{1}{(A - A_1)^2}\right]
$$

$$
=\; \frac{2}{L}\big(y_{xx}^+(x_0) - y_{xx}^-(x_0)\big) + \frac{2L^2}{A_1(A - A_1)} \;,
$$

i.e.,

$$
-\frac{1}{L}\big(y_{xx}^+(x_0) - y_{xx}^-(x_0)\big) \leqslant \frac{L^2}{A_1(A - A_1)} \;. \tag{5.3}
$$

Proposition 5.2 *Let $\gamma(\cdot, t)$ be a solution of the CSF where $\gamma(\cdot, t)$ is closed and embedded. For any $t_0 \in (0, \omega)$, either $g(\gamma(\cdot, t))$ is greater than or equal to π or it is strictly increasing in $(0, t_0)$.*

Proof: For a fixed t_0 we let $\Gamma(t_0)$ be a minimum of $g(\gamma(\cdot, t_0))$. As before, we may assume that $\Gamma(t_0)$ is a vertical line segment over $x = x_0$. By the prior discussion, $\gamma(\cdot, t)$ is locally described by the functions $y^\pm(x, t)$ for t close to t_0. Let $\Gamma(t)$ be the vertical line segment connecting $(x_0, y^-(x_0, t))$ and $(x_0, y^+(x_0, t))$. It divides the region enclosed by $\gamma(\cdot, t)$ into two regions $D_1(t)$ and $D_2(t)$ which lie on its left and right, respectively. $\big(\Gamma(t), D_1(t), D_2(t)\big)$ forms an admissible triple for $\gamma(\cdot, t)$.

The length of $\Gamma(t)$, $L(t)$, is given by $y^+(x_0, t) - y^-(x_0, t)$. Recall that $y = y^\pm$ satisfies

$$y_t = (\tan^{-1} y_x)_x \tag{5.4}$$

(see (1.3)). Let $A_1(t)$ be the area of $D_1(t)$. It follows from (5.4) that

$$\frac{dA_1}{dt} = -\pi + \tan^{-1} y_x^+(x_0, t) - \tan^{-1} y_x^-(x_0, t) . \tag{5.5}$$

By (5.1), (5.4), and (5.5) we have

$$\left.\frac{d}{dt}\right|_{t=t_0} \log L^2 \left(\frac{1}{A_1} + \frac{1}{A - A_1}\right)$$

$$= \frac{2}{L}\left(y_{xx}^+(x_0, t_0) - y_{xx}^-(x_0, t_0)\right)\left(1 + y_x^+(x_0, t_0)^2\right)^{-1}$$

$$+\left(\frac{1}{A - A_1} - \frac{1}{A}\right)\left(-\pi + 2\tan^{-1} y_x^+(x_0, t_0)\right) + 2\pi\left(\frac{1}{A - A_1} - \frac{1}{A_1}\right)$$

$$= \frac{2}{L}\left(y_{xx}^+(x_0, t_0) - y_{xx}^-(x_0, t_0)\right)\left(1 + y_x^+(x_0, t_0)^2\right)^{-1}$$

$$-2\tan^{-1} y_x^+(x_0, t_0)\left(\frac{1}{A_1} - \frac{1}{A - A_1}\right) + \pi\left[\frac{A_1^2 + (A - A_1)^2}{A_1(A - A_1)A}\right] .$$

By (5.2) and (5.3), we have

$$\left.\frac{d}{dt}\right|_{t=t_0} \log L^2 \left(\frac{1}{A_1} + \frac{1}{A - A_1}\right)$$

$$\geq -\frac{2L^2}{A_1(A - A_1)}\left(1 + y_x^+(x_0, t_0)^2\right)^{-1} - 2y_x^+(x_0, t_0)\left(\frac{1}{A_1} - \frac{1}{A - A_1}\right)$$

$$+\pi\left[\frac{A_1^2 + (A - A_1)^2}{A_1(A - A_1)A}\right]$$

$$\geq -\frac{2L^2}{A_1(A - A_1)} - \frac{L^2}{2}\left(\frac{1}{A_1} - \frac{1}{A - A_1}\right)^2 + \pi\left[\frac{A_1^2 + (A - A_1)^2}{A_1(A - A_1)A}\right]$$

$$= -\frac{L^2}{2}\left(\frac{1}{A_1} + \frac{1}{A - A_1}\right)^2 + \pi\left[\frac{A_1^2 + (A - A_1)^2}{A_1(A - A_1)A}\right]$$

$$\geq -\frac{L^2}{2}\left(\frac{1}{A_1} + \frac{1}{A - A_1}\right)^2 + \frac{\pi}{2}\left(\frac{1}{A_1} + \frac{1}{A - A_1}\right)$$

$$= \frac{1}{2}\left(\frac{1}{A_1} + \frac{1}{A - A_1}\right)\left(\pi - G(\gamma(\cdot, t_0))\right).$$

When $G(\gamma(\cdot, t_0)) < \pi$, this inequality shows that $g(\gamma(\cdot, t))$ is strictly increasing in $[t_0 - \delta, t_0]$ for some $\delta > 0$. By repeating the same argument to any minimum of $g(\gamma(\cdot, t_0 - \delta))$, we can trace back in time and show that $g(\gamma(\cdot, t))$ is strictly increasing in $(0, t_0]$. $\square$

We shall need a similar isoperimetric ratio where the line segments lie in the exterior of γ. Let's assume that γ is an embedded, closed curve which is non-convex and contained in the unit disk. Let C be the circle $\{x^2 + y^2 = 16\}$. We may consider the collection of all line segments which belong to the region D bounded between C and γ and whose endpoints touch γ. As before, each such line segment Γ divides D into D_1 and D_2, and we may form the triple (Γ, D_1, D_2). The isoperimetric quantity $G'(\Gamma)$ and $g'(\gamma)$ can be defined in a similar way. It is still true that any minimizing Γ intersects γ transversally at the endpoints and its interior belongs to D. Furthermore, the first and second variation formulas (5.1)–(5.3) hold for Γ.

Proposition 5.3 *Let γ be a solution of the CSF where $\gamma(\cdot, t)$ is embedded, closed, and non-convex for all t. Then, for any fixed $t_0 \in (0, \omega)$, either $g'(\gamma(\cdot, t_0)) \geq \pi/2$ or $g'(\gamma(\cdot, t))$ is strictly increasing in $(0, t_0]$.*

Proof: As in the proof of Proposition 5.2, we assume the minimum of $g(\gamma(\cdot, t_0))$, $\Gamma(t_0)$, is vertical over $x = x_0$, and $\gamma(\cdot, t)$ is described by $y^{\pm}(\cdot, t)$ near the endpoints. Let $A(t)$ be the area of the region $D(t)$

bounded by $\gamma(\cdot, t)$ and C, and $A_1(t)$ the area of the sub-region $D_1(t)$ lying on the left of $\Gamma(t)$. We have, by (1.19),

$$\frac{dA}{dt} = 2\pi , \tag{5.6}$$

and (5.5) holds. By (5.1)–(5.3), (5.5), and (5.6) we have

$$\frac{d}{dt}\bigg|_{t=t_0} \log L^2 \left(\frac{1}{A_1} + \frac{1}{A - A_1} \right)$$

$$= \frac{2}{L}(y_{xx}^+(x_0, t_0) - y_{xx}^-(x_0, t_0))(1 + y_x^+(x_0, t_0)^2)^{-1}+$$

$$\left(\frac{1}{A - A_1} - \frac{1}{A_1} \right)\left(-\pi + 2\tan^{-1} y_x^+(x_0, t_0) \right) + \left(\frac{1}{A} - \frac{1}{A - A_1} \right)\frac{dA}{dt} \Bigg]$$

$$\geqslant -\frac{2L^2}{A_1(A - A_1)} - \frac{L^2}{2}\left(\frac{1}{A_1} - \frac{1}{A - A_1} \right)^2 + \pi\left(\frac{1}{A} - \frac{1}{A - A_1} \right)$$

$$+ 2\pi\left(\frac{1}{A} - \frac{1}{A - A_1} \right)$$

$$= -\frac{L^2}{2}\left(\frac{1}{A_1} + \frac{1}{A - A_1} \right)^2 + \pi\frac{(A - A_1)^2 - 3A_1^2}{A_1(A - A_1)A}.$$

Since $\gamma(\cdot, t)$ is contained inside the unit disk, $A_1(t) < \pi$ and $A(t) - A_1(t) \geqslant 15\pi$. Therefore,

$$\pi\frac{(A - A_1)^2 - 3A^2}{A_1(A - A_1)A} \geqslant \frac{\pi}{2}\frac{A_1^2 + (A - A_1)^2}{A_1(A - A_1)A}$$

$$\geqslant \frac{\pi}{4}\left(\frac{1}{A_1} + \frac{1}{A - A_1} \right),$$

which implies

$$\frac{d}{dt} \log L^2 \left(\frac{1}{A_1} + \frac{1}{A - A_1} \right)$$

$$\geqslant \frac{1}{2}\left(\frac{1}{A_1} + \frac{1}{A - A_1} \right)\left(\frac{\pi}{2} - g'(\gamma(\cdot, t_0)) \right).$$

Now we can finish the proof of the proposition following the last step in the proof of Proposition 5.2.   □

5.2 Limits of the rescaled flow

By (1.18) and Proposition 1.2, the curvature of the maximal solution of the CSF for an immersed closed γ_0 becomes unbounded as $t \uparrow \omega$. Consequently, we can find a sequence $\{\gamma(p_n, t_n)\}$, $t_n \uparrow \omega$ and $p_n \in S^1$ such that

$$|k(p, t)| \leqslant |k(p_n, t_n)|, \qquad \forall \ (p, t) \in S^1 \times [0, t_n] \ .$$

We call such a sequence an **essential blow-up sequence**. Notice that, by (1.17), the curvature may tend to negative infinity.

We rescale the flow by setting

$$\gamma_n(p, t) = \frac{\gamma(p_n + p, t_n + \varepsilon_n^2 t) - \gamma(p_n, t_n)}{\varepsilon_n},$$

for $t \in [-t_n/\varepsilon_n^2, (\omega - t_n)/\varepsilon_n^2)$, where $\varepsilon_n = |k(p_n, t_n)|^{-1}$. So for each n, γ_n solves the curve shortening flow, $\gamma_n(0, 0) = (0, 0)$, and its curvature satisfies $|k_n| \leqslant |k_n(0, 0)| = 1$ for $t \in [-t_n/\varepsilon_n^2, 0]$. From now on, we shall assume each γ_n is parametrized by arc-length. Since the limit curve may not be closed, it is convenient to assume each γ_n is defined on the real line as a periodic map. By Remark 1.3 and the Ascoli-Arzela theorem, we can extract a subsequence of γ_n which converges to a solution of the CSF in every compact subset of $\mathbb{R} \times (-\infty, 0]$. The limit solution γ_∞, defined in $\mathbb{R} \times (-\infty, 0]$, is either closed, or unbounded and complete for each t. Moreover, $|k_\infty(p, t)| \leqslant |k_\infty(0, 0)| = 1$. We have:

Proposition 5.4 (i) $\gamma_\infty(\cdot, t)$ is uniformly convex in $(-\infty, 0]$, and (ii) if γ_∞ is embedded and unbounded, its total curvature must be equal to π.

Lemma 5.5 *The total absolute curvature of any CSF for closed curves is non-increasing in time.*

Proof: In each positive ε, we have

$$\frac{d}{dt}\int_{\gamma(\cdot,t)}(\varepsilon^2 + k^2(\cdot,t))^{\frac{1}{2}}ds$$

$$= \int_{\gamma(\cdot,t)}k(\varepsilon^2+k^2)^{-\frac{1}{2}}\left(\frac{\partial^2 k}{\partial s^2}+k^3\right)ds - \int_{\gamma(\cdot,t)}k^2(\varepsilon^2+k^2)^{\frac{1}{2}}ds$$

$$= -\int_{\gamma(\cdot,t)}\varepsilon^2(\varepsilon^2+k^2)^{-\frac{3}{2}}\left(\frac{\partial k}{\partial s}\right)^2 ds - \int_{\gamma(\cdot,t)}\varepsilon^2 k^2(\varepsilon^2+k^2)^{-\frac{1}{2}}ds$$

$$\leqslant \quad 0.$$

Letting $\varepsilon \downarrow 0$, we obtain the desired result.   □

Lemma 5.6 *Any inflection point of $\gamma_\infty(\cdot,t)$ must be degenerate.*

Proof: As above, we have

$$\frac{d}{dt}\int_{\gamma_n(\cdot,t)}(\varepsilon^2 + k_n^2(\cdot,t))^{\frac{1}{2}}ds \tag{5.7}$$

$$= -\int_{\gamma_n(\cdot,t)}\varepsilon^2(\varepsilon^2+k_n^2)^{-\frac{3}{2}}\left(\frac{\partial k_n}{\partial s}\right)^2 ds - \int_{\gamma_n(\cdot,t)}\varepsilon^2 k_n^2(\varepsilon^2+k_n^2)^{-\frac{1}{2}}ds.$$

Suppose at some t_0, $\gamma_\infty(s_0, t_0)$ is a nondegenerate inflection point. As γ_n tends to γ_∞ smoothly on every compact set, for sufficiently large n and small $\delta > 0$, the following holds: for each $t \in [t_0 - \delta, t_0]$ and large n, there exists $s_n(t)$ near s_0 such that

$$k_n(s_n(t), t) = 0$$

and

$$\left|\frac{\partial k_n}{\partial s}(s_n(t), t)\right| \geq \frac{1}{2}\left|\frac{\partial k_\infty}{\partial s}(s_0, t_0)\right| > 0.$$

Now, the first term on the right-hand side of (5.7) can be estimated as follows:

$$-\int_{\gamma_n} \varepsilon^2(\varepsilon^2 + k_n^2)^{-\frac{3}{2}}\left(\frac{\partial k_n}{\partial s}\right)^2 ds$$

$$\leqslant -\int_{s_n(t)-\varepsilon}^{s_n(t)+\varepsilon} C_1\varepsilon^2[\varepsilon^2 + C_2(s - s_n(t))^2]^{-\frac{3}{2}}\,ds$$

$$\leqslant -C_1'\int_{-C_2'\varepsilon}^{C_2'\varepsilon} \varepsilon^2(\varepsilon^2 + s^2)^{-\frac{3}{2}}\,ds$$

$$= \frac{-2C_1'C_2'}{(1 + C_2'^2)^{\frac{1}{2}}},$$

where the positive constants C_1, C_2, C_1', and C_2' depend on $\left|\frac{\partial k_\infty}{\partial s}(s_0, t_0)\right|$ only.

By integrating (5.7) from $t_0 - \delta$ to t_0, we have

$$\int_{\gamma_n(\cdot,t_0)} (\varepsilon^2 + k_n^2)^{\frac{1}{2}}\,ds \leqslant \int_{\gamma_n(\cdot,t_0-\delta)} (\varepsilon^2 + k_n^2)^{\frac{1}{2}}\,ds - \frac{2C_1'C_2'\delta}{(1 + C_2'^2)^{\frac{1}{2}}}.$$

Letting $\varepsilon \downarrow 0$, we arrive at

$$\int_{\gamma_n(\cdot,t_n)} |k_n|\,ds \leqslant \int_{\gamma_n(\cdot,t_0-\delta)} |k_n|\,ds - \frac{2C_1'C_2'\delta}{(1 + C_2'^2)^{\frac{1}{2}}}.$$

Back to the unscaled flow, it means that

$$\int_{\gamma(\cdot,t_n+\varepsilon_n^2 t_0)} |k|\,ds - \int_{\gamma(\cdot,t_n+\varepsilon_n^2(t_0-\delta))} |k|\,ds \leqslant -\frac{2C_1'C_2'\delta}{(1 + C_2'^2)^{\frac{1}{2}}}.$$

However, since the total absolute curvature of γ is non-increasing in t, the left-hand side of this inequality tends to zero as $n \to \infty$, and the contradiction holds. We conclude that γ_∞ cannot have any non-degenerate inflection point. $\square$

Proof of Proposition 5.4 (i) By Lemma 5.6, γ_∞ does not have any non-degenerate inflection points. Were it not uniformly convex, we can find some t, p, and q such that $k_\infty(p, t)k_\infty(q, t) < 0$ and there is a point r between p and q satisfying $k_\infty(r, t) = 0$. By applying the Sturm oscillation theorem to the equation satisfied by k_∞, we infer that, for $t' > t$, $t' - t$ small, γ_∞ has a non-degenerate inflection point. This is in conflict with Lemma 5.6. Hence, $\gamma_\infty(\cdot, t)$ must be uniformly convex for all t.

(ii) Since now γ_∞ is unbounded and embedded, its total curvature is less than or equal to π. Let's show that the former is impossible. To prove this, we introduce coordinates so that, for a fixed t_0, $\gamma(\cdot, t_0)$ is the graph of a convex function $U(x, t_0)$, $x \in \mathbb{R}$, with $|U_x| \to a$ as $|x| \to \infty$. By comparing $\gamma(\cdot, t)$ with vertical lines which are stationary solutions of the CSF, every $\gamma(\cdot, t)$ is the graph of a smooth function $U(x, t)$. Next, by comparing $\gamma(\cdot, t)$ with grim reapers which move steadily upward or downward, we infer that the speed of $\gamma(\cdot, t)$ is bounded by a constant. In particular, it means that $|U_x(x, t)| \to a$ as $|x| \to \infty$ for all t. By differentiating (5.4), we see that the function $w = U_x$ is a positive solution to the following uniformly parabolic equation in divergence form,

$$w_t = \left(\frac{w_x}{1 + w^2} \right)_x ,$$

and hence it is constant by Moser's Harnack inequality (Theorem 6.28 in [88]). It means $\gamma_\infty(\cdot, t)$ is a horizontal line for all t, contradicting with its uniform convexity. So, the total curvature of γ_∞ is always equal to π. The proof of Proposition 5.4 is completed.

Now, we can prove the Grayson convexity theorem. If γ_∞ is bounded, it follows from Proposition 5.4 that, for sufficiently large

n, γ_n is uniformly convex too. Scaling back, we conclude that $\gamma(\cdot, t)$ is uniformly convex for all t close to ω.

If γ_∞ is unbounded, by Proposition 5.4, it can be expressed as the graph of a family of convex functions $U(x, t)$, $x \in \mathbb{R}$ or $(-a, a)$, satisfying $U_x(x, t) \to \pm\infty$ as $x \to \pm\infty$ (or $\pm a$). We shall show that this will lead to a contradiction.

For any $\varepsilon > 0$, we can find x_ε such that

$$U(x, 0) \geqslant \frac{1}{\varepsilon} x_\varepsilon, \qquad x \in (-a, a), |x| \geqslant x_\varepsilon,$$

where we have assumed $U(0, 0) = 0$. Consider the horizontal line segment $\Gamma = \{y = \frac{1}{\varepsilon} x_\varepsilon\} \cap \{y > U(x, 0)\}$. It is clear that the length L of Γ is not greater than $2x_\varepsilon$, and, by convexity, the area A bounded between Γ and $\gamma_\infty(\cdot, 0)$ is not less than $2^{-1}L \cdot x_\varepsilon/\varepsilon$. For sufficiently large n, the length L_n of $\Gamma_n = \{y = x_\varepsilon/\varepsilon\} \cap \{y > \gamma_n(\cdot, 0)\}$ and the area A_n of the region D'_n bounded between Γ_n and $\gamma_n(\cdot, 0)$ are arbitrarily close to L and A, respectively. Therefore, when Γ_n belongs to D_n (the region enclosed by $\gamma_n(\cdot, 0)$), the triple $(\Gamma_n, D'_n, D_n \setminus D'_n)$ is admissible, and we have

$$g(\gamma_n(\cdot, 0)) \leqslant G(\Gamma_n)$$

$$< 2L^2 \left(\frac{1}{\frac{1}{2}L \cdot \frac{x_\varepsilon}{\varepsilon}} + \frac{1}{C_n} \right),$$

where C_n tends to ∞ as $n \to \infty$. By the Euclidean and scaling invariance of $g(\gamma)$, it shows that $g(\gamma(\cdot, t))$ could be arbitrarily small for some t close to ω. But this is impossible by Proposition 5.2. On the other hand, if Γ_n belongs to the exterior of $\gamma_n(\cdot, 0)$, we can then use Proposition 5.3 instead of Proposition 5.2 to draw the same contradiction. (Without loss of generality, we may assume $\gamma(\cdot, t)$ is contained in the unit disk.) Hence, γ_∞ must be unbounded. We have finished the proof of the theorem.

5.3   Classification of singularities

First of all, we note that the blow-up rate of the curvature for the
CSF has a universal lower bound. For, it follows from $(1.17)'$ that

$$\frac{dk_{\max}}{dt} \leqslant k_{\max}^3 .$$

By integrating this inequality from t to ω, we obtain

$$k_{\max}(t) \geqslant \left[2(\omega - t)\right]^{-1/2} .$$

With a similar estimate on $k_{\min}(t)$, we conclude

$$|k|_{\max}(t) \geqslant \left[2(\omega - t)\right]^{-1/2} . \tag{5.8}$$

Let $\gamma(\cdot, t)$ be a CSF for closed, immersed curves. By Theorem 6.4, there are finitely many singularities as t approaches ω. A
point Q in $\mathbb{R}^2$ in called a **singularity** for the CSF $\gamma(\cdot, t)$ if there
exists $\{(p_j, t_j)\}$, $p_j \in S^1$, $t_j \uparrow \omega$, such that $\gamma(p_j, t_j) \longrightarrow Q$ and
$|k|(p_j, t_j) \longrightarrow \infty$ as $j \longrightarrow \infty$. The singularity is **of type I** if there
exist a constant C and a neighborhood $\mathcal{U}$ of Q which is disjoint from
other singularities such that

$$\sup_{\mathcal{U}} |k|(\cdot, t) \sqrt{2(\omega - t)} \leqslant C \tag{5.9}$$

for all $t \in [0, \omega)$. It is **of type II** if (5.9) does not hold.

When Q is a type I–singularity, it is natural to rescale it by
setting

$$\widetilde{\gamma}(\cdot, \tau) = \left[2(\omega - t)\right]^{-1/2} \left(\gamma(\cdot, t) - Q\right) ,$$

where $2\tau = -\log(\omega - t)$. When $\gamma(\cdot, t)$ is embedded, this normalization coincides with the one we used in Chapter 3. Notice from
(5.8), (5.9), and parabolic regularity, that the curvature of $\widetilde{\gamma}$ is uniformly pinched between two positive constants and its derivatives are
uniformly bounded for all τ.

Theorem 5.7 *Let $\gamma(\cdot, t)$ be a CSF for closed, immersed curves. Suppose that all singularities are of type I. Then, each sequence $\{\tilde{\gamma}(\cdot, \tau_j)\}$ contains a subsequence converging to a self-similar solution of the CSF contracting to the origin as $\tau_j \longrightarrow \infty$.*

It follows from this theorem the flow eventually becomes convex and shrinks to a point.

The proof of Theorem 5.7 is based on a monotonicity formula. Let the "normalized backward heat kernel" be

$$\rho(X) = e^{-|X|^2/2} \; , \; X \in \mathbb{R}^2 \; .$$

We define

$$\mathcal{M}(\tau) = \int_{\tilde{\gamma}(\cdot, \tau)} \rho\big(\tilde{\gamma}(\cdot, \tau)\big) d\tilde{s} \; ,$$

where $\tilde{s}$ is the arc-length parameter of $\tilde{\gamma}$.

Lemma 5.8

$$\frac{d}{d\tau} \mathcal{M}(\tau) = - \int_{\tilde{\gamma}(\cdot, \tau)} \rho \big|\tilde{k} + \langle \tilde{\gamma}, n \rangle\big|^2 d\tilde{s} \; .$$

Proof: $\tilde{\gamma}$ and its arc-length $\tilde{s}$ satisfy the equations

$$\tilde{\gamma}_t = \tilde{k} n + \tilde{\gamma} \; ,$$

and

$$\tilde{s}_\tau = (-\tilde{k}^2 + 1)\tilde{s} \; .$$

Therefore,

$$\begin{aligned}
\frac{d}{d\tau} \mathcal{M}(\tau) &= \int \rho(-\tilde{k}^2 + 1) - \int \rho\langle \tilde{\gamma}, \tilde{k}n + \tilde{\gamma}\rangle d\tilde{s} \\
&= - \int \rho\big(|\tilde{k}n + \tilde{\gamma}|^2 - \langle \tilde{\gamma}, \tilde{\gamma}_{\tilde{s}}\rangle\big) d\tilde{s} \\
&= - \int \rho|\tilde{k} + \langle \tilde{\gamma}, n \rangle|^2 d\tilde{s},
\end{aligned}$$

after a direct computation.   □

   Now, by (5.9),

$$|\gamma(s,t) - Q| \leqslant \int_t^\omega |k| dt$$
$$\leqslant C\sqrt{\omega - t} \ .$$

This means

$$|\widetilde{\gamma}(s,\tau)| \leqslant C \ ,$$

that is, $\widetilde{\gamma}$ is contained in the disk D_C for all τ. Since all derivatives of $\widetilde{k}$ are under control, it follows from Lemma 5.8 that

$$\lim_{\tau \longrightarrow \infty} \int \rho |\widetilde{k} + \langle \widetilde{\gamma}, n \rangle|^2 d\widetilde{s} = 0 \ .$$

Thus, any sequence $\{\widetilde{\gamma}(\cdot, t_j)\}$ has a subsequence converging to a closed curve γ^* solving

$$k^* + \langle \gamma^*, n \rangle = 0 \ .$$

By reversing the orientation of γ^*, if necessary, k^* is positive somewhere. We may introduce the support function of γ^*, h^*, and then this equation becomes

$$h^* = k^* \ .$$

So, γ^* is a contracting self-similar solution. All self-similar solutions of the CSF have been classified in Chapter 2. Since γ^* is closed, it must be one of the Abresch-Langer curves. The proof of Theorem 5.7 is completed.

   Next, consider a type II singularity Q. Suppose that there is an essential blow-up sequence converging to it in some neighborhood of Q, $\mathcal{U}$, which contains no other singularities. It is appropriate to use the normalization described in the previous section. Adapting the same notation, the normalized sequence γ_n subconverges to a limit flow $\gamma_\infty(\cdot, t)$, $t \in (-\infty, 0]$.

Theorem 5.9 *Let Q be a type II singularity. Suppose that there exists an essential blow-up sequence converging to Q in some neighborhood $\mathcal{U}$. Then γ_∞ is a grim reaper.*

We point out that the assumption of the existence of an essential blow-up sequence converging to Q is not necessary. In fact, according to Theorem 6.4, singularities of the limit curves γ^* as $t \uparrow w$ are finite. Hence, one can apply the blow-up argument in any fixed, sufficiently small neighborhood of each singularity, where an essential blow-up sequence always exists.

Proof: First of all, by the definition of a type II singularity and the definition of γ_n, we know that γ_n subconverges to γ_∞ in every compact subset of $\mathbb{R}$. Hence, γ_∞ is a CSF which exists for all positive and negative t. (A solution which exists for all time is called an eternal solution.) In the previous section, we have shown that it is uniformly convex. We claim that it has no self-intersection. For, the rate of decrease in the area of a loop is equal to the difference of the normal angles at the meeting point, which is at least π. Hence, γ_∞ cannot exist for all time if it has a loop. Now, as before, one can show that the total curvature of γ_∞ must be equal to π.

We write $\gamma = \gamma_\infty$ for simplicity. We claim that a convex eternal solution must be a travelling wave. For, let $f = k_t$ and $g = 2(t - t_0)f + k$ and parameterize the flow by the normal angle. Then,

$$f_t = k^2(f_{\theta\theta} + f) + \frac{2f^2}{k}$$

and

$$g_t = k^2 g_{\theta\theta} + \left(k^2 + \frac{2f}{k}\right)g .$$

When γ is closed, the strong maximum principle implies that $g > 0$ for all $t > t_0$. By approximating γ by a closed, convex flow, one sees that it continues to hold for the unbounded γ. Letting t_0 go back to $-\infty$, we conclude that $k_t \geqslant 0$ for all $t \in \mathbb{R}$.

Now, when γ is closed, consider its entropy:

$$\mathcal{E}(t) = \int \log k(\theta, t)d\theta .$$

As before, we have

$$\frac{d^2}{dt^2}\mathcal{E}(t) = 2\int \frac{k_t^2}{k^2}d\theta$$

$$\geqslant 2\left(\frac{d\mathcal{E}(t)}{dt}\right)^2 .$$

So

$$\frac{d\mathcal{E}}{dt}(t) \leqslant \frac{\pi}{2(T-t)}$$

for all $T > t$. By an approximation argument, this inequality still holds for unbounded γ. Letting $T \longrightarrow \infty$, we conclude that $k_t = 0$, i.e., k is a travelling wave for the CSF.   □

Example 5.10 Type II singularities may occur even when the flow shrinks to a point. For example, let γ_0 be a figure-eight which is symmetric with respect to the x-axis and the y-axis with exactly one inflection point located at the origin. It is easy to see that the flow exists until the area enclosed by each leaf becomes zero. By symmetry, the limit curve is either a point or a line segment. The latter is excluded by the strong maximum principle. So, the flow shrinks to a point is finite time. Now, since there are no Abresch-Langer curves with two leaves, the singularity must be of type II. In general, singularities of figure-eight were studied in Grayson [67], where it is shown that the difference of the enclosed area of the leaves is constant in time. When the areas are equal but the leaves are not symmetric, it is not known whether the flow shrinks to a point or not. In any case, the singularities are of type II.

Example 5.11 A cardoid is a closed curve with winding number 2, symmetric with respect to the x-axis, and on which the curvature has one maximum, one minimum, and no other critical points in between these. It is clear that the inside closed loop will shrink to a point before the outside one does, and, hence, forms a type II singularity. A delicate analysis, carried out in Angenent-Velazquez [18], shows that

$$k_{\max}(t) = \left(1 + o(1)\right) \left[\frac{\log|\log(\omega - t)|}{\omega - t} \right]^{1/2}$$

and the asymptotic shape near the cusp at ω looks like

$$y(x) = \left(\frac{\pi}{4} + o(1) \right) \frac{x}{\log|\log x|}.$$

Notes

The proof of the Grayson convexity theorem presented here is different from the original proof and is taken from Hamilton [72]. This approach by rescaling, blowing up and analyzing the singularities has been used in many problems in geometric analysis, including nonlinear heat equations (Giga-Kohn [65]), harmonic heat flows (Struwe [105]), Ricci flows (Hamilton [73]), and mean curvature flows (Huisken and Sinestrari [79]).

The results in Section 5.3 are mainly based on Altschuler [3].

Similar results can be found in Angenent [15] in the convex case. The monotonicity property, Lemma 5.8, which also holds for the mean curvature flow, is due to Huisken [77]. For monotonicity properties in other geometric problems, see [65], [105], and [73].

Isoperimetric ratios

In [72], Hamilton gave two isoperimetric ratios, both of which are non-decreasing along the CSF. We have used the first one and complemented it with Proposition 5.3 due to Zhu [113]. The second one, which is more precise, can be described as follows. Let Γ be any curve which divides the region D enclosed by $\gamma(\cdot, t)$ into two regions D_1 and D_2. Consider the shortest curve $\overline{\Gamma}$ which divides the disk $\overline{D}$, whose area is equal to $|D|$, into two regions of area equal to $|D_1|$ and $|D_2|$, respectively. Define

$$H(\gamma) = \frac{\text{length } |\Gamma|}{\text{length } |\overline{\Gamma}|}$$

and

$$h(\gamma) = \inf_{\Gamma} H(\gamma) \ .$$

Then, $h(\gamma(\cdot, t))$ increases along the flow. More recently, two more scaling-invariant geometric ratios were introduced by Huisken [78]. Let ℓ and d be, respectively, the intrinsic and extrinsic distance: $S^1 \times S^1 \times [0, w) \to \mathbb{R}$ on $\gamma(\cdot, t)$. Define

$$I(\gamma) \;=\; \frac{d}{\ell} \ ,$$

$$J(\gamma) \;=\; d \Big/ \frac{L}{\pi} \sin \frac{\ell \pi}{L} \quad (L \text{ is the perimeter of } \gamma(\cdot, t)) \ ,$$

$$i(\gamma) = \inf_{(p,q)} I(\gamma) \ ,$$

and

$$j(\gamma) = \inf_{(p,q)} J(\gamma) \ .$$

Then, $i(\gamma(\cdot, t))$ and $j(\gamma(\cdot, t))$ increase along the flow. (For $i(\gamma(\cdot, t))$ we need to impose the assumption that the infimum is attained in the interior.) Both h and j can be used in place of g to prove the Grayson convexity theorem. We shall discuss $i(\gamma)$, which is especially effective for non-closed curves, in some detail in Chapter 8.

Boundary value problems for the CSF. The ratio $i(\gamma)$ is very useful in the study of various boundary value problems of the CSF ([78]). Combined with the blow-up argument, it can be used to control the curvature of the flow away from the boundary. For example, the following result is valid:

Theorem *Let $\gamma_0 : I \rightarrow \mathbb{R}$ be an embedded curve lying between the strip $\{a_1 < x < a_2\}$ with endpoints P_i on the vertical line $x = a_i$, $i = 1, 2$. Then, the CSF has a unique solution $\gamma(\cdot, t)$, $\in [0, \infty)$, satisfying $\gamma(a_i, t) = P_i$, $i = 1, 2$. Moreover, $\gamma(\cdot, t)$ tends to the line segment connecting P_1 and P_2 as $t \rightarrow \infty$.*

See Huisken [76], Polden [95], and Stahl [102] for further results on the boundary value problems of the CSF.

The CSF for complete, noncompact curves. In Ecker-Huisken [47], it was proved that the mean curvature flow for any entire graph has a solution for all time. This is rather striking because the behaviour of the initial hypersurface at infinity does not affect solvability. In [95], long-time existence is established for any initial curve whose ends are asymptotic to some semi-infinite lines. In Chou-Zhu [34], we prove the following general result.

Theorem. *Let γ_0 be any embedded curve which divides the plane into two regions of infinite area. Then, the CSF has a solution for all time.*

An example has been constructed to show that ω is finite when one of the regions has finite area. In the same paper, it is also proved that the solution is unique when the ends of γ_0 are graphs over some semi-infinite lines. Long-time behaviour, such as convergence to an expanding self-similar solution or a grim reaper, can be found in [46] and [95].

The long-time existence of the curvature-eikonal flow for any initial entire graph is established in Chou-Kwong [30].

Motion of curves in space. Flows in space may be written as

$$\frac{\partial \gamma}{\partial t} = F\boldsymbol{n} + G\boldsymbol{t} + H\boldsymbol{b} \qquad (5.10)$$

where $\boldsymbol{b}$ is the binormal of $\gamma(\cdot, t)$ and F, G and H depend the curvature and the torsion. The CSF ($H = G = 0$ and $F = k$) was studied by Altschuler-Grayson [4] where long time existence was established for a special class of curves called "ramps". However, the purpose of [4] is to use spatial evolution as a mean to study the long time behavior of the CSF in the plane, especially after the formation of singularity. A different approach which is based on expressing the flow as a weakly parabolic system was developed in Deckelnick [41] in $\mathbb{R}^n$ ($n \geqslant 2$). For the level-set approach one can consult Ambrosio-Soner [7].

It is worthwhile to point out that the general flow (5.10) sometimes arises in applications, for example in the dynamics of vortex filaments in fluid dynamics ($F = G = 0$ and $H = k$, Hasimoto [74] and [91]) and in the dynamics of scroll waves in excitable media (Keener-Tyson [83]).

Chapter 6

A Class of Non-convex Anisotropic Flows

In this chapter, we continue the study of flows for non-convex curves. Let Φ and Ψ be two smooth, 2π-period functions of the tangent angle satisfying

$$\Phi(\theta) > 0 \tag{6.1}$$

and

$$\Phi(\theta + \pi) = \Phi(\theta) \ , \ \ \Psi(\theta + \pi) = -\Psi(\theta) \ . \tag{6.2}$$

We consider the Cauchy problem for

$$\frac{\partial \gamma}{\partial t} = (\Phi k + \Psi)\boldsymbol{n} \ , \tag{6.3}$$

where the initial curve γ_0 is a smooth, embedded closed curve. This flow may be regarded as the linear case for the general flow (1.2), where F is uniformly parabolic and symmetric. Remember that the condition (6.2) means that F is symmetric. Without this condition, embeddedness may not be preserved under the flow. When γ_0 is convex, we have shown in §3.2 that the flow also preserves convexity and it shrinks to a point where ω is finite. In this chapter we shall show that the Grayson convexity theorem holds for (6.3).

Theorem 6.1 *Consider the Cauchy problem for (6.3) where (6.1) and (6.2) hold. For any embedded closed γ_0, there exists a unique* *solution to the Cauchy problem in $[0, \omega)$ where ω is finite. Moreover, for each t in $[0, \omega)$, $\gamma(\cdot, t)$ is embedded closed and there exists some t_0 such that $\gamma(\cdot, t)$ become uniformly convex in $[t_0, \omega)$.*

Notice that Theorem 6.1 reduces to the Grayson convexity theorem when $\Phi \equiv 1$ and $\Psi \equiv 0$. The method we are going to use to prove this theorem is completely different from that of the previous chapter. Here we follow the elementary and geometric approach initiated by Grayson [66] and subsequently developed by Angenent [14]. It is based on the construction of suitable foliations and then uses them to bound the curvature of the flow. We shall see that the Sturm oscillation theorem plays a crucial role in this approach.

6.1   The decrease in total absolute curvature

Consider the Cauchy problem for (6.3). From now on, we shall always assume (6.1) and (6.2) are in force and γ_0 is an embedded closed curve. According to Propositions 1.2 and 1.4, it admits a unique smooth solution $\gamma(\cdot, t)$ in a maximal interval $[0, \omega)$. When (6.2) holds and γ_0 is embedded, each $\gamma(\cdot, t)$ is embedded.

First of all, let's show that ω is finite. In fact, by (1.18), the length of $\gamma(\cdot, t)$ satisfies

$$\frac{dL}{dt} = -\int_{\gamma} \Phi(\theta) k^2 ds .$$

By the Hölder inequality,

$$L^2(t) \leqslant L^2(0) - 8\pi^2 \Phi_{\min} t .$$

So, ω is bounded above by a constant depending only on the length of the initial curve.

Next, we examine the inflection points of the flow. By (1.16), the curvature of $\gamma(\cdot, t)$, $k(\cdot, t)$, satisfies the equation

$$k_t = (\Phi k + \Psi)_{ss} + k^2 (\Phi k + \Psi) ,$$

where $s = s(t)$ is the arc-length parameter of $\gamma(\cdot, t)$. Using $\partial \theta / \partial s = k$, this equation can be rewritten in the form

$$k_t = \Phi s^{-2} k_{pp} + b k_p + c k ,$$

where the coefficients b and c are smooth. Since $\gamma(\cdot, t)$ is a closed curve and so its curvature never vanishes identically, it follows from Fact 7 in §1.2 that the zero set of $k(\cdot, t)$, i.e., the inflection points of $\gamma(\cdot, t)$, is finite for every $t \in (0, \omega)$. Moreover, the number of inflection points drops precisely at those instants t when $\gamma(\cdot, t)$ has a degenerate inflection point, and all these instants form a discrete set in $(0, \omega)$.

Let's assume that the flow is always nonconvex, for otherwise Theorem 6.1 holds trivially and there is nothing to prove. Thus, there are at least two inflection points on the solution curve for each time. Without loss of generality, we may assume the number of inflection points on $\gamma(\cdot, t)$ is constant in $(0, \omega)$. Denote the set of inflection points on $\gamma(\cdot, t)$ by $S(t)$. There exist $N \geqslant 2$ and smooth functions $p_1(t), \cdots, p_N(t)$ to S^1 such that, for each $j = 1, \cdots, N$, the arc $c^j(t) \equiv \gamma(\cdot, t)\big|_{[p_j(t), p_{j+1}(t)]}$ $(p_{N+1} \equiv p_1)$ has non-vanishing curvature except at their endpoints. Recall that the tangent angle at $\gamma(p, t)$ is determined as follows. First, we assign its value at a fixed $\gamma(p_0, t_0)$. Then, we extend it to all $\gamma(p, t)$ by continuity. As a result, any two choices of tangent angles differ by a constant multiple of 2π everywhere. Let $\theta_j(t)$ be the tangent angle at $\gamma(p_j(t), t)$, and let $[\theta_j(t), \theta_{j+1}(t)]$ or $[\theta_{j+1}(t), \theta_j(t)]$ be the range of the tangent angle of a convex or concave arc. Clearly, the tangent angle along an arc is strictly increasing or decreasing depending on whether the arc is uniformly convex or concave.

We claim that θ_j is strictly increasing and θ_{j+1} is strictly decreasing in time along a convex arc c_j. This implies that the interval $[\theta_j(t), \theta_{j+1}(t)]$ is strictly nesting in time. To prove the claim, let t_0 be any fixed time and let $\gamma(p_0, t_0)$ be in $S(t_0)$. By representing the solution as a local graph $(x, u(x, t))$ over, say, the x-axis, the function u satisfies

$$u_t = \sqrt{1 + u_x^2} \left(\Phi(\theta)k + \Psi(\theta) \right) ,$$

where $\gamma(p_0, t_0) = (x_0, u(x_0, t_0))$, $\theta = \tan^{-1} u_x$. By differentiating this equation, we have

$$\theta_t = \cos^2 \theta \left(\Phi \theta_x \right)_x + (\cos \theta \Psi_\theta + \Psi \sin \theta) \theta_x .$$

Since $\gamma(p_0, t_0)$ is a non-degenerate inflection point, the tangent angle at $\gamma(p_0, t_0)$, $\theta(p_0, t_0)$ is either a strict local maximum or minimum. For some small $\delta > 0$, we have either

$$\theta > \theta(p_0, t_0)$$

or

$$\theta < \theta(p_0, t_0)$$

on the parabolic boundary of $(x_0 - \delta, x_0 + \delta) \times (t_0, t_0 + \delta)$. By the strong maximum principle, we conclude that either

$$\min_{(x_0 - \delta, x_0 + \delta)} \theta(\cdot, t) > \theta(p_0, t_0) \;\; \text{or}$$

$$\max_{(x_0 - \delta, x_0 + \delta)} \theta(\cdot, t) < \theta(p_0, t_0) ,$$

for $t \in (t_0, t_0 + \delta)$. Noticing that $\theta(p_0, t_0)$ is a local strict maximum (resp. minimum) when it is equal to θ_{j+1} (resp. θ_j), our claim follows. By a similar argument, one can show that the range of the tangent angle of a concave arc is also strictly nested.

Finally, we note that the total absolute curvature of $\gamma(\cdot, t)$, when it has no degenerate inflection points, is given by

$$\sum_{j=1}^{N} \left| \theta_{j+1}(t) - \theta_j(t) \right| .$$

Therefore, it decreases strictly as long as $\gamma(\cdot, t)$ is nonconvex. Summing up, we have proved the following proposition.

Proposition 6.2 *There exists a unique solution to the Cauchy problem for (6.3) in $(0, \omega)$ where ω is finite. When $\gamma(\cdot, t)$ is non-convex in $(0, \omega)$, the range of tangent angles of each convex\concave arc of $\gamma(\cdot, t)$ is strictly nesting. As a result, the total absolute curvature of $\gamma(\cdot, t)$ is strictly decreasing.*

6.2 The existence of a limit curve

The total absolute curvature of a closed convex curve is always equal to 2π. In view of Proposition 6.2, the flow (6.3) has a tendency towards convexity. We shall first use this fact to show that a limit curve exists when t approaches ω. To achieve this goal, we need to "foliate" the solution curves.

By a **foliation** we mean a smooth diffeomorphism $\mathcal{F}$ from $I \times [0, 1]$, where I is either a closed interval or a closed arc in S^1, to $\mathbb{R}^2$ such that the **leaf** $\mathcal{F}(\mu, \cdot)$ is a smooth curve without self-intersections for each μ. We shall denote a foliation by $\mathcal{F}_\mu(\cdot)$ or by $\mathcal{F}(\mu, \cdot)$.

Given a foliation $\mathcal{F}$, we can solve (6.3) using each leaf as the initial curve. Granted solvability, we obtain in this way a family of time-varying foliations $\mathcal{F}_\mu(\cdot, t)$, $t \in [t_1, t_2]$. To make use of the foliation, we actually need to "bend" it to obtain two foliations which lie on its left and right, respectively.

For each small $\delta > 0$, let's postulate the existence of two foliations $\mathcal{F}^+$ and $\mathcal{F}^-$ of (6.3) which satisfy

(H$_1$) $|\mathcal{F}_\mu^\pm(p,t) - \mathcal{F}_\mu(p,t)| \leqslant \delta$, $(p,\mu) \in I \times [0,1]$, $t \in [t_1, t_2]$; and

(H$_2$) Any two different leaves from $\mathcal{F}$, $\mathcal{F}^+$, and $\mathcal{F}^-$ are either disjoint or meet transversally at exactly one point. At any intersection point of a leaf of $\mathcal{F}^+(\cdot,t)$ and a leaf of $\mathcal{F}^-(\cdot,t)$, there passes a leaf of $\mathcal{F}(\cdot,t)$ such that the leaf of $\mathcal{F}^+(\cdot,t)$ (resp. the leaf of $\mathcal{F}^-(\cdot,t)$) lies on its right (resp. left). The angle between the leaves has a uniform positive lower bound.

By an **evolving arc**, we mean a smooth map $\beta : \Omega \to \mathbb{R}^2$ where $\Omega = \bigcup\{[p_1(t), p_2(t)] \times \{t\} : t \in [0,T)\}$ and $[p_1(t), p_2(t)]$ is a continuously changing arc of S^1 such that $\beta(\cdot,t)$ solves (6.3) in $[0,T)$ and its endpoints are always separated:

$$\inf_t | \beta(p_1(t),t) - \beta(p_2(t),t) | > 0 .$$

Proposition 6.3 *Let $\beta(\cdot,t), t \in [0,T)$, be an evolving arc of (6.3). Suppose that there are three foliations described as above and satisfying*

   (i) $\beta(\cdot,0)$ is transversal to $\mathcal{F}_\mu(\cdot,0)$ and $\mathcal{F}_\mu^\pm(\cdot,0)$ whenever they meet; and

   (ii) for each t, the endpoints $\beta(\cdot,t)$ are disjoint from $\mathcal{F}_\mu(\cdot,t)$ and $\mathcal{F}_\mu^\pm(\cdot,t)$, and the endpoints of $\mathcal{F}_\mu(\cdot,t)$ and $\mathcal{F}_\mu^\pm(\cdot,t)$ are disjoint from $\beta(\cdot,t)$.

Let $Q(t) = \bigcup\{\mathcal{F}_\mu(\cdot,t) : \mu \in [0,1]\}$ and let D be a region compactly supported inside $Q(t)$ for all t. Then, there exists a constant C depending on $\mathcal{F}$, $\mathcal{F}^\pm$, $\beta(\cdot,0)$, and D so that the curvature of $\{\beta(p,t) : p \in \Omega \cap D\}$ is bounded by C (provided $\Omega \bigcap D$ is non-empty).

Proof: For each $t \in [0,T)$, $\mathcal{F}$ defines a diffeomorphism from $[0,1] \times [a,b]$ to $Q(t)$. By (ii), we may assume each $\beta(p,t)$ enters $\mathcal{F}$ at $\mathcal{F}(0,t)$ and leaves $Q(t)$ through $\mathcal{F}(1,t)$. As long as it is transveral to $\mathcal{F}_\mu(\cdot,t)$,

p and μ are in one-to-one correspondence, and, by the inverse function theorem,

$$\mathcal{F}(\mu, y(\mu, t), t) = \beta(p(\mu, t), t)$$

for some smooth y. By differentiating this equation,

$$\frac{\partial \mathcal{F}}{\partial \mu} + \frac{\partial \mathcal{F}}{\partial y} y_\mu = \beta_p \frac{\partial p}{\partial \mu} \,,$$

$$\frac{\partial^2 \mathcal{F}}{\partial \mu^2} + 2 \frac{\partial^2 \mathcal{F}}{\partial \mu \partial y} y_\mu + \frac{\partial^2 \mathcal{F}}{\partial y^2} y_\mu^2 + \frac{\partial \mathcal{F}}{\partial y} y_{\mu\mu}$$

$$= \beta_{pp} \left(\frac{\partial p}{\partial \mu} \right)^2 + \beta_p \frac{\partial^2 p}{\partial \mu^2} \,,$$

and

$$\frac{\partial \mathcal{F}}{\partial t} + \frac{\partial \mathcal{F}}{\partial y} y_t = \beta_p \frac{\partial p}{\partial t} + \beta_t \,.$$

It follows that

$$\mathbf{t} = \left(\frac{\partial \mathcal{F}}{\partial \mu} + \frac{\partial \mathcal{F}}{\partial y} y_\mu \right) \Big/ \left| \frac{\partial \mathcal{F}}{\partial \mu} + \frac{\partial \mathcal{F}}{\partial y} y_\mu \right| \,,$$

$$k = \left\langle \mathbf{n}, \frac{\partial \mathcal{F}}{\partial y} y_{\mu\mu} + \frac{\partial^2 \mathcal{F}}{\partial \mu^2} + 2 \frac{\partial^2 \mathcal{F}}{\partial \mu \partial y} y_\mu + \frac{\partial^2 \mathcal{F}}{\partial y^2} y_\mu^2 \right\rangle \Big/ \left| \frac{\partial \mathcal{F}}{\partial \mu} \right|^2 \,,$$

and

$$y_t = \Phi(\theta) \left| \frac{\partial \mathcal{F}}{\partial \mu} \right|^{-2} y_{\mu\mu} + B(y, y_\mu) \,, \tag{6.4}$$

where B depends on Φ, y_μ, and the partial derivatives of $\mathcal{F}$ (evaluated
at (μ, y, t)).

On the other hand, denote the preimages of $\mathcal{F}^+(\cdot, t)$ and $\mathcal{F}^-(\cdot, t)$
under $\mathcal{F}(\cdot, t)$ by $\widehat{\mathcal{F}}^+(\cdot, t)$ and $\widehat{\mathcal{F}}^-(\cdot, t)$, respectively. Let $\widehat{D}$ be a region
compactly supported inside $(0, 1) \times (a, b)$ such that it contains all
preimages of D under $\mathcal{F}(\cdot, t)$, $t \in [0, T]$. By choosing δ sufficiently

small in (H_1), $\widehat{\mathcal{F}}^+(\cdot, t)$ and $\widehat{\mathcal{F}}^-(\cdot, t)$ foliate $\widehat{D}$ for each t. Moreover, by (H_2), each leaf in $\widehat{\mathcal{F}}^+$ (resp. $\widehat{\mathcal{F}}^-$) is the graph of a strictly increasing (resp. strictly decreasing) function z^+ (resp. z^-) whose gradients are bounded by a constant C_0.

In view of the formula for k, it suffices to bound y_μ and $y_{\mu\mu}$. Since $z^\pm$ also satisfy (6.4) and (ii) holds, the Sturm oscillation theorem gives

$$-C_0 \leqslant \frac{\partial z^-}{\partial \mu} \leqslant \frac{\partial y}{\partial \mu} \leqslant \frac{\partial z^+}{\partial \mu} \leqslant C_0 \ .$$

In particular, it implies that $\beta(\cdot, t)$ is always transversal to $\mathcal{F}^\pm(\cdot, t)$ in D and so y satisfies (6.4) whenever $(\mu, y(\mu, t))$ belongs to $\widehat{D}$. Now, as in the proof of Proposition 1.9, the function y_μ satisfies a uniformly parabolic equation in divergence form. By Theorem 3.1 in Chapter 5 of [86], we conclude that $y_{\mu\mu}$ is bounded in any compact subset of $\widehat{D}$. $\square$

Theorem 6.4 *Suppose $\gamma(\cdot, t)$ is a maximal solution of (6.3) in $[0, \omega)$ where γ_0 is immersed and closed. Then, as $t \uparrow \omega$, the curve $\gamma(\cdot, t)$ converges to a limit set γ^* in the Hausdorff metric. The limit set γ^* is the image of a Lipschitz continuous map. When γ^* is not a point, there exist finitely many points $\{Q_1, \cdots, Q_m\}$ on γ^* such that $\gamma^* \backslash \{Q_1, \cdots, Q_m\}$ consists of smooth curves. Away from these points $\gamma(\cdot, t)$ converge smoothly to γ^*.*

Proof: We begin by reparametrizing the curve $\gamma(\cdot, t)$ so that $|\gamma_p(p, t)|$, $p \in S^1$, is a constant depending on t only. We may assume that $\gamma(\cdot, t)$ does not shrink to a point. Then, there are two positive numbers L_1 and L_2 so that the length of $\gamma(\cdot, t)$ is always bounded between L_1 and L_2. So, $|\gamma_p(\cdot, t)|$ is bounded between L_1^{-1} and L_2^{-1}. On the other hand, by Proposition 1.10, we know that the image of $\gamma(\cdot, t), t \in (0, \omega)$, is contained in a bounded set. By the Ascoli-Arzela theorem, we can extract a subsequence for $\gamma(\cdot, t)$ which converges uniformly to some Lipschitz continuous map γ^*. By Proposition 1.10, it is clear that $\gamma(\cdot, t)$ converges to γ^* in the Hausdorff metric, that is, for every $\varepsilon > 0$, there exists t_0 such that $\gamma(\cdot, t)$ is contained in the ε-neighborhood of γ^* for all $t \in [t_0, \omega)$.

The curvature of $\gamma(p, t)$, $k(p, t)$, induces a metric $|k(p, t)| dp$ on the unit circle. Denote the push-forward of this measure under $\gamma(\cdot, t)$ by $\boldsymbol{K}_t$. Then $\{\boldsymbol{K}_t\}$ is a family of uniformly bounded Borel measures on $\mathbb{R}^2$. We can select a subsequence $\boldsymbol{K}_{t_n}$ which converges to some Borel measure $\boldsymbol{K}$ weakly. This limit measure can be decomposed into atoms and a continuous part,

$$ \boldsymbol{K} = \boldsymbol{K}_c + \sum_{j=1} K_j \delta_{Q_j} , $$

where $\boldsymbol{K}_c(\{P\}) = 0$ for any point P, $\{Q_j\}$ is at most a countable set of points, arranged in the way $K_1 \geqslant K_2 \geqslant K_3 \geqslant \cdots > 0$. We let m be the integer for which $K_m \geqslant \pi$ and $K_{m+1} < \pi$.

We shall show that $\{Q_j\}$ contains m many points. In fact, let P be any point in $\gamma^* \backslash \{Q_1, \cdots, Q_m\}$. We can find α, $\varepsilon > 0$, and n_0 such that $\boldsymbol{K}_{t_n}(D_\varepsilon(P)) < \pi - \alpha$ for all $n \geqslant n_0$. The preimage of $D_\varepsilon(P)$ under any one of these $\gamma_n = \gamma(\cdot, t_n)$ is a countable disjoint union of intervals in S^1. Since the length of $\gamma(\cdot, t)$ is uniformly bounded, the number of these intervals whose images also intersect $D_{\varepsilon/2}(P)$ has a finite upper bound. By passing to a subsequence, if necessary, we may assume there are exactly N arcs of $\gamma_n(\cdot)$ contained in $D_\varepsilon(P)$ which also intersect $D_{\varepsilon/2}(P)$ for all $n \geqslant n_0$.

The ranges of tangent angles of these components are intervals. At each t_n, the sum of the lengths of these intervals does not exceed $\pi - \alpha$. By rotating the coordinates and passing to a subsequence, again, if necessary, we may assume the intervals $[\pi/2 - \alpha/(3N), \pi/2 + \alpha/(3N)]$ and $[3\pi/2 - \alpha(3N), 3\pi/2 + \alpha/(3N)]$ lie in the complement of the range of tangent angles for $D_\varepsilon(P) \cap \gamma_n(\cdot)$ for all $n \geqslant n_0$. In other words, the arcs are graphs of some Lipschitz continuous functions $y_{n,1}(x), \cdots, y_{n,N}(x)$ whose Lipschitz constants are bounded by $\cot\alpha/(3N)$. Since $\{\gamma(\cdot, t)\}$ tends to γ^* in the Hausdorff metric, the functions $\{y_{n,i}(x)\}$ converge to γ^* uniformly as $n \to \infty$. Taking P to be the origin, we can find $\xi, \eta > 0$, $\xi \geqslant \eta \tan\alpha/(4N)$ such that the two lines $\ell_\pm = [-\xi, \xi] \times \{\pm\eta\}$ lie inside $D_{\varepsilon/2}(P)$ and are disjoint from $\gamma(\cdot, t)$ and γ^* for all $t \geqslant t_{n_0}$.

According to the basic theory of ODEs, there is a unique stationary solution of (6.3) with $(0, \pm\eta)$ as endpoints for small ε. Denote this solution arc by ω_0 and translate it along the x-axis to obtain a foliation $\mathcal{F}$ consisting of stationary arcs $\omega_x, x \in [-\xi, \xi]$. When ε_0 is sufficiently small, we may assume the tangent angle of every leaf lies in $[\pi/2 - \alpha/(400N), \pi/2 + \alpha/(400N)]$. Consequently, for each $t_n \geqslant t_{n_0}$, $\gamma(\cdot, t)$ intersects ω_x, at most, N times. By passing to a subsequence again, we may assume the number of intersection, is always N. Similarly, we may consider stationary arcs ω_0^+ (resp. ω_0^-) connecting $(0, -\eta)$ and $(\eta \tan\alpha/(100N), \eta))$ (resp. $(-\eta \tan\alpha/(100N), \eta))$. For small ε, the ranges of their tangent angles are contained in $[\pi/2 - \alpha/(100N) - \alpha/(400N), \pi/2 - \alpha/(100N) + \alpha/(400N))$ (resp. $[\pi/2 + \alpha/(100N) - \alpha/(400N), \pi/2 + \alpha/(100N) + \alpha/(400N)])$. Translate them to obtain two foliations $\mathcal{F}^+$ and $\mathcal{F}^-$.

Now, we can find a small disk D containing the origin that is covered by $\mathcal{F}, \mathcal{F}^+$, and $\mathcal{F}^-$. By applying Proposition 6.3 to each component of $\gamma(\cdot, t)$, $t \geqslant t_{n_0}$, and the foliations $\mathcal{F}^\pm$, we conclude that the curvature of $\gamma(\cdot, t)$ is uniformly bounded in D near ω. Letting $t \uparrow \omega$, this implies that γ^* cannot have a singularity at Q. The contradiction holds. □

6.3 Shrinking to a point

In this section, we show that the flow $\gamma(\cdot,t)$ of (6.3) shrinks to a point as $t \uparrow \omega$.

Let $\beta(p,t)$, $p \in [p_1(t), p_2(t)]$, $t \in (t_1, t_2)$, be an embedded evolving arc given by $\beta(\cdot,t) = \gamma(\cdot,t) \cap D$, where D is a disk, and $\partial\beta(\cdot,t)$ consists of two distinct points on the boundary of D. For any point P on $\beta(\cdot,t)$, let $\theta(P,t)$ be the tangent angle of $\beta(\cdot,t)$ at P. We consider

$$\Theta(t;\beta) = \max\left\{ |\theta(P,t) - \theta(P',t)| \; : \; P, P' \text{ any points on } \beta(\cdot,t) \right\} .$$

So, $\Theta(t;\beta)$ measures the maximal change in tangent angles along a convex or concave arc of $\beta(\cdot,t)$.

Lemma 6.5 *Let $\beta(\cdot,t)$ be an evolving arc of an embedded closed flow of (6.3) described as above. Suppose that (i) the curvature of $\beta(\cdot,t)$ is uniformly bounded in a δ-neighborhood of the parabolic boundary of $\bigcup\{[p_1(t), p_2(t)] \times \{t\} : t \in [t_1, t_2)\}$ by C_0, and (ii) there exists $C_1 > 0$ such that*

$$\sup_t \Theta(t;\beta) \leqslant C_1 . \tag{6.5}$$

Then, the curvature of $\beta(\cdot,t)$, $t \in [t_1, t_2)$, is uniformly bounded in $[p_1(t), p_2(t)] \times [t_1, t_2)$ by a constant depending on C_0 and C_1.

Before proving this lemma, let's first show how to use it to deduce the following result.

Theorem 6.6 *Let $\gamma(\cdot,t)$ be an embedded solution of (6.3). Then, it shrinks to a point as $t \uparrow \omega$.*

Proof: Suppose on the contrary that $\gamma(\cdot,t)$ does not shrink to a point. According to Theorem 5.4, we can find a limit curve γ^* and

$\{Q_1, \cdots, Q_m\} \subseteq \gamma^*, m \geqslant 1$, such that the curvature of $\gamma(\cdot, t)$ becomes unbounded near $Q_j, j = 1, \cdots, m$. For each $Q = Q_j$, we can find a small $\varepsilon > 0$ such that $\beta(\cdot, t) = \gamma(\cdot, t) \cap D_\varepsilon(Q)$ satisfies that hypotheses of Lemma 6.5. Notice that $\Theta(t, \beta)$ is always uniformly bounded by the total absolute curvature of $\gamma(\cdot, t)$, which is strictly decreasing along the flow. By Lemma 6.5, we arrive at a contradictory conclusion—the curvature of $\beta(\cdot, t)$ is uniformly bounded. So $\gamma(\cdot, t)$ must shrink to a point. $\square$

We shall prove Lemma 6.5 by an induction argument on C_1 in two steps.

Step 1 Assume that the hypotheses of Lemma 6.5 hold and, in addition, $C_1 \leqslant \pi - \alpha$ for some $\alpha > 0$. We prove the lemma.

Step 2 Assuming that Lemma 6.5 holds whenever $C_1 \leqslant \dfrac{n\pi}{4} - \alpha$ for some $n \geqslant 4$, we show that it continues to hold for all $C_1 \leqslant \dfrac{n+1}{4}\pi - \alpha$.

Combining Step 1 and Step 2, it follows from induction that Lemma 6.5 holds for any positive constant C_1.

Proof of Step 1 Suppose on the contrary there is an arc $\beta(\cdot, t)$ satisfying the hypotheses of the lemma, but its curvature blows up as $t \uparrow t_2$. Let $\{\gamma(p_n, t_n)\}$ be an essential blow-up sequence inside D (see §5.2). Following the blow-up procedure there, we can find a limit curve $\beta_\infty : (-\infty, 0] \times \mathbb{R} \longrightarrow \mathbb{R}^2$ satisfying

$$\frac{\partial \beta_\infty}{\partial t} = \Phi(\theta) k_\infty \boldsymbol{n} , \quad k_\infty(0, 0) = 1 .$$

Notice that, although $\Phi \equiv 1$ and $\Psi \equiv 0$ in the blow-up argument described in §5.2, the same argument applies without any change to the present situation. The curvature bound near the parabolic boundary guarantees that β_∞ is a complete, unbounded curve. Furthermore, its total curvature is less than or equal to $\pi - \alpha$.

Now, we can represent $\beta_\infty(\cdot, t)$ as the graph of a smooth function

v in $\mathbb{R} \times (-\infty, 0]$ whose derivatives satisfy

$$|v_x| \leqslant \tan^{-1}\left(\frac{\pi - \alpha}{2}\right) , \quad v_x(0,0) = 0 , \quad v_{xx}(0,0) = 1 ,$$

for all t, and it satisfies the equation

$$v_t = \Phi(\theta)\frac{v_{xx}}{1 + v_x^2} ,$$

where $\theta = \tan^{-1} v_x + \theta_0$, $\theta_0 \in [0, 2\pi)$. The constant θ_0 results from the change of coordinates which puts β_∞ into a graph over the x-axis. Now, as before, we observe that the non-negative function $w = v_x$ satisfies the uniformly parabolic equation

$$w_t = \left(\Phi(\theta)\frac{w_x}{1 + w^2}\right)_x .$$

By Moser's Harnack inequality ([88]), we conclude that w is a constant. The contradiction holds.

Proof of Step 2 First of all, we observe that it suffices to show Step 2 holds for $t_2 = \omega$ and t_1, a fixed time very close to ω. Lemma 6.5 holds trivially if $t_2 < \omega$. Second, in view of Theorem 6.4, it suffices to show Step 2 for disks of the form $D_\varepsilon(Q)$ where $Q \in \{Q_1, \cdots, Q_m\}$ and ε is sufficiently small.

In the following, we shall establish the validity of Step 2 on $D_\varepsilon(Q) \times [\bar{t}, \omega)$ where ε and $\bar{t}$ will be chosen sufficiently close to 0 and ω, respectively. Let $Q = Q_j$ be fixed and ε_0 is so small that $D_{\varepsilon_0}(Q)$ contains no singularities other than Q, and $\gamma(\cdot, t) \cap D_{\varepsilon_0}(Q)$ is connected for all $t \in [\bar{t}, \omega)$. Let $\beta^* = \gamma^* \cap D_{\varepsilon_0}(Q)$ and $\beta_j^* = \beta^* \cap (D_{\varepsilon_j}(Q) \setminus D_{\varepsilon_{j+1}}(Q))$, where $\varepsilon_j = (100)^{-j}\varepsilon_0$. Since $\beta(\cdot, t)$ tends to β^* smoothly away from Q and the total absolute curvature of $\gamma(\cdot, t)$ is uniformly bounded, there exists ℓ such that

$$\int_{\beta_j^*} |k|ds < \frac{\pi}{200} , \quad \text{for all } j \geqslant \ell .$$

We choose $\bar{t}$ such that

$$\int_{\beta_\ell(t)} |k| ds < \frac{\pi}{100} , \quad \text{for all } t \in [\bar{t}, \omega) .$$

(We shall restrict $\bar{t}$ further as we proceed.) From now on, we write $\varepsilon_1 = \varepsilon_\ell$. Observing that the total absolute curvature along $\beta_\ell(t)$ is very small and $D_{\varepsilon_{\ell+1}}(Q)$ is very small compared to $D_{\varepsilon_1}(Q)$, the two arcs comprising $\beta_\ell(t)$ are virtually straight rays emitting from the origin, and are almost perpendicular to the boundary of $D_{\varepsilon_1}(Q)$.

We would like to determine a time-varying foliation $\mathcal{F}_\mu(\cdot, t)$ which meets the following requirement: for all $t \in [\bar{t}, \omega)$,

(i) the endpoints of $\beta'(\cdot, t) \equiv \beta(\cdot, t) \cap D_{\varepsilon_1}(Q)$, $P_1(t)$ and $P_2(t)$, lie outside $\mathcal{F}_\mu(\cdot, t)$,

(ii) $\mathcal{F}_\mu(\cdot, t)$ covers a fixed small neighborhood of Q,

(iii) $\beta'(\cdot, t)$ passes through the leaves of $\mathcal{F}_\mu(\cdot, t)$ without tangency, and,

(iv) for each μ, $\Theta\big(t; \mathcal{F}_\mu(\cdot, t) \cap D_{\varepsilon_1}(Q)\big) \leqslant \dfrac{n\pi}{4} - \alpha$.

Let $\xi = \pi/100$. We rotate the coordinate axes so that the tangent angle $\theta(\cdot, t)$ of $\beta'(\cdot, t)$ satisfies

$$\liminf_{t \uparrow \omega} \{\theta(P, t) : P \in \beta'(\cdot, t)\} = 0 . \tag{6.6}$$

Then (iii) and (iv) will be satisfied if $\mathcal{F}_\mu$ is defined in such a way that its tangent angle ϕ satisfies

$$\phi \in \Big[-2\xi, (n+1)\frac{\pi}{4} - \alpha - \frac{\pi}{2} + \xi \Big] \tag{6.7}$$

and, whenever $\mathcal{F}_\mu$ meets $\beta(\cdot, t)$,

$$\theta - \phi \in \Big[\frac{\xi}{2}, \frac{\pi}{2} + \frac{\xi}{2} \Big] . \tag{6.8}$$

In our construction, each leaf of $\mathcal{F}_\mu(\cdot, t)$ will be an embedded closed curve. Then, (i) and (ii) will also be satisfied if the boundary leaves $\mathcal{F}_1(\cdot, t)$ and $\mathcal{F}_2(\cdot, t)$ separate $P_1(t)$ from Q and $P_2(t)$ from Q, respectively.

Let $P_i^*(i = 1, 2)$ be the endpoints of $\beta^* \cap D_{\varepsilon_1}(Q)$. According to our choice of ε_1, for a number δ which is very small compared to ε_1, the circle $C_{2\delta}(P_i^*)$ intersects $\beta'(\cdot, t)$ at exactly one point X_i, $i = 1, 2$. Let $\ell_i = \overline{A_i B_i}$ be the chord of $D_{\varepsilon_1}(Q)$ formed by the straight line passing through X_i and pointing at the direction $(\cos \phi_i, \sin \phi_i)$, where $\phi_i = \max\{\theta(X_i, \bar{t}) - \pi/2, -\xi\}, i = 1, 2$. We assume that $(\cos \phi_i, \sin \phi_i)$ points outward at B_i and inward at A_i. Define a vector field $\mathbf{v_0} = (\cos \phi, \sin \phi)$ along $\beta'(\cdot, \bar{t})$ by setting

$$\phi = \max\left\{\theta - \frac{\pi}{2}, \ -\xi\right\}, \tag{6.9}$$

and $\phi = \phi_i$, along ℓ_i for $i = 1, 2$. Notice that $\mathbf{v_0}$ is transversal along $\beta'(\cdot, t)$ and it satisfies (6.7) and (6.8). The vector field is Lipschitz continuous. We replace it by a smooth approximation and still denote it by $\mathbf{v_0}$. By (6.6) and (6.9), one can check that the two chords ℓ_1 and ℓ_2 are disjoint. (In fact, the choice of $D_{\varepsilon_1}(Q)$ and $D_{\varepsilon_{l+1}}(Q)$ shows that $\beta'(\cdot, t)$ is almost perpendicular to the boundary of $D_{\varepsilon_1}(Q)$. It is shown in [92] that the angle between $\beta'(\cdot, t)$ and $\partial D_{\varepsilon_1}(Q)$ at $P_i(t)$, $i = 1, 2$, is bounded between $\pi/2 \pm \pi/50$. When $\phi = \theta - \pi/2$, the chord ℓ_1 or ℓ_2 is very short. On the other hand, if $\phi = -\xi$ at both endpoints, large portions of the chords ℓ_1 and ℓ_2 lie inside different components of $D_{\varepsilon_1}(Q) \setminus \beta'(\cdot, t)$.) Also, together with the arcs $\overset{\frown}{A_1 A_2}$ and $\overset{\frown}{B_1 B_2}$, they bound a sub-region of $D_{\varepsilon_1}(Q)$, R, such that $D_{\varepsilon_1/100}(Q)$ is contained in R. Since $\mathbf{v_0}$ points inward at A_1 and A_2, and outward at B_1 and B_2, we can define a smooth $\mathbf{v_0}$ along the arc $\overset{\frown}{A_1 A_2}$ (resp. $\overset{\frown}{B_1 B_2}$) so that it always points inward (resp. outward), and (6.7) continues to hold.

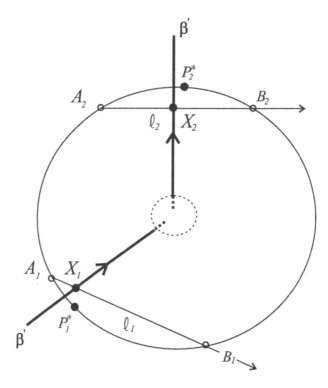

Figure 6.1

The large disk in this figure is $D_{\varepsilon_1}(Q)$ and the small one is $D_{\varepsilon_1/100}(Q)$.

By the Tietze Extension Theorem, we can extend $\boldsymbol{v_0}$ to a vector field (still denoted by) $\boldsymbol{v_0}$ on $\overline{R}$ satisfying (6.7). Since $\boldsymbol{v_0}$ is smooth, we may assume without loss of generality that the extended $\boldsymbol{v_0}$ is also smooth. Any integral curve of $\boldsymbol{v_0}$ is transversal to $\beta'(\cdot, t)$, and its endpoints lie on the arcs $\overparen{A_1 A_2}$ and $\overparen{B_1 B_2}$. We may extend each integral curve outside R by smoothly attaching a line segment to each endpoint. Let $\ell_i' = A_i' B_i'$ be the extended line segment. With a necessary restriction on δ, we may assume $\overparen{A_1' A_2'}$ and $\overparen{B_1' B_2'}$ lie on $C_{\varepsilon_1 + 2\delta}(Q)$. In this way, we have found a foliation of extended integral curves on the region R' bounded between $\ell_1', \ell_2', \overparen{A_1' A_2'}$, and $\overparen{B_1' B_2'}$. Finally, we extend each integral curve outside $D_{\varepsilon_1 + 2\delta}(Q)$ by closing it up with a large arc to obtain a foliation $\mathcal{F}$ which consists of embedded closed curves with $\ell_1' \subseteq \mathcal{F}_0$ and $\ell_2' \subseteq \mathcal{F}_1$. $\mathcal{F}_0$ separates $D_{2\delta'}(P_1^*)$ from $D_{2\delta'}(Q)$ and $D_{2\delta'}(P_2^*)$, and $\mathcal{F}_1$ separates $D_{2\delta'}(P_2^*)$ from $D_{2\delta'}(Q)$ and $D_{2\delta'}(P_1^*)$. Here, $\delta' = C\delta$ and $C < 1$ is a constant determined by ξ.

Let $\mathcal{F}(\cdot, t)$ be the foliation obtained by solving (6.3) using each leaf $\mathcal{F}_\mu$ as the initial curve at $\bar{t}$. We claim that there is an *a priori* uniform bound on the curvature of the leaves of this foliation outside $D_{\varepsilon_1 + 3\delta/2}(Q)$ in $[\bar{t}, \bar{t} + \rho]$ for some $\rho > 0$. For, first of all, there is a uniform bound on the curvature of the leaves outside $D_{\varepsilon_1 + 3\delta/2}(Q)$ at time $\bar{t}$. Therefore, for any point X on $\mathcal{F}_\mu(\cdot, t)$, one can find a small square S_ℓ centered at X, whose side ℓ is less than $\delta/2$, such that $\mathcal{F}_\mu(\cdot, \bar{t}) \cap S_\ell$ is the graph of a function over one side of S_ℓ. When ℓ is sufficiently small, the tangent angles of this graph are nearly constant. Using Proposition 1.10, the existence of stationary arcs of (6.3) transversal to the graph, the Sturm oscillation theorem, and parabolic regularity, we know that there is some $\rho > 0$ such that the curvature of $\mathcal{F}_\mu(\cdot, t) \cap S_{\ell/2}$ is uniformly bounded for all $t \in [\bar{t}, \bar{t} + \rho]$. Since X could be any point on $\mathcal{F}_\mu(\cdot, \bar{t}) \setminus D_{\varepsilon_1 + 3\delta/2}(Q)$, the curvature of $\mathcal{F}_\mu(\cdot, t)$ outside $D_{\varepsilon_1 + 3\delta/2}(Q)$ is uniformly bounded in $[\bar{t}, \omega]$, provided we choose $\bar{t}$ such that $\bar{t} + \rho \geqslant \omega$. Next, consider the curvature of $\mathcal{F}_\mu(\cdot, t)$ inside $D_{\varepsilon_1 + 2\delta}(Q)$. At $t = \bar{t}$,

$$\Theta\left(\bar{t}; \mathcal{F}_\mu(\cdot, \bar{t}) \cap D_{\varepsilon_1 + 2\delta}(Q)\right) \leqslant \frac{n\pi}{4} - \frac{\pi}{4} + 3\xi - \alpha \, ,$$

by construction. Since convex\concave arcs are nesting and the curvature of $\mathcal{F}_\mu(\cdot, t)$ near the ends is uniformly bounded,

$$\Theta\left(t; \mathcal{F}_\mu(\cdot, t) \cap D_{\varepsilon_1 + 2\delta}(Q)\right) \leqslant \frac{n\pi}{4} - \alpha$$

for all $t \in [\bar{t}, \omega]$ when $\bar{t}$ is close to ω. By induction hypothesis, the curvature is also bounded in $D_{\varepsilon_1 + 2\delta}(Q)$. By Proposition 1.2, the foliation $\mathcal{F}_\mu(\cdot, t)$ exists in $[\bar{t}, \omega]$.

We still have to make sure that the endpoints of $\beta'(\cdot, t)$ never touch the foliation $\mathcal{F}_\mu(\cdot, t)$. Consider the circles $C_{2\delta'}(P_1^*), C_{2\delta'}(P_2^*)$, and $C_{2\delta'}(Q)$. Using them as initial curves starting at $\bar{t}$, we solve (6.3) to obtain three flows. When δ' is sufficiently small, the flows are shrinking and we can find $\bar{t}$ close to ω such that they still contain the disks $D_{\delta'}(P_1^*), D_{\delta'}(P_2^*)$, and $D_{\delta'}(Q)$, respectively for $t \in [\bar{t}, \bar{t}+\omega]$. Since $P_i(t)$ tends to $P_i^*, i = 1, 2$, as $t \uparrow \omega$. We may assume that the endpoints of $\beta'(\cdot, t)$ are contained inside $D_{\delta'}(P_1^*)$ and $D_{\delta'}(P_2^*)$ during the same time interval.

So, finally we have constructed a foliation and found a $\bar{t}$ such that our requirements (i)—(iv) are satisfied. It is straightforward to bend the foliation to the left and right to obtain two foliations satisfying the assumptions in Proposition 6.3. By this proposition, the curvature of $\beta'(\cdot, t)$ is uniformly bounded in $D_{\delta'}(Q)$ for all t in $[\bar{t}, \omega)$. Hence, Step 2 holds.

6.4   A whisker lemma

So far, we have shown that the flow $\gamma(\cdot, t)$ shrinks to a point as $t \uparrow \omega$. Further, since the number of inflection points does not increase, we may assume that it is always equal to N in $[0, \omega)$. We shall continue to assume $\gamma(\cdot, t)$ is nonconvex and, hence, $N \geqslant 2$. We can decompose $\gamma(\cdot, t)$ into a union of convex\concave evolving arcs $c_1(t), \cdots, c_N(t)$ whose total absolute curvature is either less than $\pi - \delta$ or greater than π for some $\delta > 0$.

Lemma 6.7 *The curvature on an arc $c_j(t)$ is uniformly bounded if the total absolute curvature of $c_j(t)$ is less than $\pi - \delta$.*

Proof: Let $c(t)$ be such an arc. We can use the tangent angle θ to parametrize it. The equation for the curvature, $k(\theta, t)$, is given by

$$k_t = k^2 \left[(\Phi k + \Psi)_{\theta\theta} + (\Phi k + \Psi) \right] . \tag{6.10}$$

Without loss of generality, let's assume the tangent angles along $c(0)$ lie in $[\delta/2, \pi - \delta/2]$. Then, the tangent angles of $c(t)$ lie in the same interval for all subsequent time t. Equation (6.10) admits two stationary solutions

$$k^{\pm} = \frac{1}{\Phi} \left[\frac{\pm \left(|\Psi|_{\max} + \Phi_{\max} |k|_{\max}(0) \right)}{\sin \delta/2} \sin \theta - \Psi \right] .$$

At $t = 0$, $k^-(\theta) \leqslant k(\theta, 0) \leqslant k^+(\theta)$, for all θ in $[\delta/2, \pi - \delta/2]$. As k^+ is non-negative and k^- is non-positive in $[\delta/2, \pi - \delta/2] \times [0, \infty)$, by the comparison principle, we have

$$|k(\theta, t)| \leqslant \Phi_{\min}^{-1} \left[\frac{|\Psi|_{\max} + \Phi_{\max} k_{\max}(0)}{\sin \delta/2} + |\Psi|_{\max} \right] .$$

$\square$

Let γ be a closed curve. We call an arc α of γ a **nice arc** (with respect to a coordinate system) if

(i) the curvature on γ is negative and bounded away from zero, and

(ii) the tangents at the endpoints of γ are horizontal.

Since the tangent angles (or normal angles) of each $c_j(t)$ are strictly nesting in time, for any given nice arc $\alpha(t_0)$ of $\gamma(\cdot, t_0)$, there exists a family of continuously changing nice arcs $\alpha(t)$, of $\gamma(\cdot, t)$, $t \leqslant t_0$, connecting $\alpha(t_0)$ to some nice arc $\alpha(0)$.

Let $\alpha(t)$ be such a family of nice arcs of $\gamma(\cdot, t), t \in [0, t_0)$. We define the $\boldsymbol{\delta}$-**whisker of** $\alpha(t)$ to be the set $\{X = \alpha(t) + \mu e : \mu \in (0, \delta)$ and e is the unit horizontal tangent pointing to the interior of $\gamma(\cdot, t)\}$.

Lemma 6.8 *There exists $\delta > 0$ depending on $\gamma(\cdot, 0)$ such that, for any nice arc $\alpha(t_0)$ on $\gamma(\cdot, t_0), t_0 \in [0, \omega)$, its δ-whisker is contained in the interior of $\gamma(\cdot, t_0)$.*

Proof: Let $\beta(t) = \gamma(\cdot, t) \backslash \alpha(t)$ be the complement of $\alpha(t)$. We first connect $\alpha(t_0)$ through nice arcs to $\alpha(0)$. Then, we define

$$d(t) = \min \left\{ \left| \beta^1(t) - \alpha^1(t) \right| \right\},$$

where $\alpha(t) = \left(\alpha^1(t), \alpha^2(t) \right)$ is a point on $\alpha(t)$ and $\beta(t) = \left(\beta^1(t), \beta^2(t) \right)$ is a point on $\beta(t)$ satisfying $\beta^2(t) = \alpha^2(t)$. We shall prove the lemma by showing that $d(t)$ is increasing in $[0, t_0]$. To this end, let's take $e = (1, 0)$. (The other case $e = (-1, 0)$ can be handled similarly.) Let $\widetilde{\beta}(t)$ be the portion of $\beta(t)$ that is bounded between the two horizontal tangents lines passing the endpoints of $\alpha(t)$. Every horizontal line segment with endpoints on $\alpha(t)$ and $\widetilde{\beta}(t)$ lies entirely inside $\gamma(\cdot, t)$. Then

$$d(t) = \min \left\{ \left| \widetilde{\beta}^1(t) - \alpha^1(t) \right| \right\},$$

where $\alpha(t) = \left(\alpha^1(t), \alpha^2(t) \right)$ is a point on $\alpha(t)$ and $\widetilde{\beta}(t) = \left(\widetilde{\beta}^1(t), \widetilde{\beta}^2(t) \right)$ is a point on $\widetilde{\beta}(t)$ satisfying $\widetilde{\beta}^2(t) = \alpha^2(t)$. Geometrically speaking, $d(t)$ is the minimal distance covered when we translate $\alpha(t)$ to its right horizontally until it first hits $\widetilde{\beta}(t)$.

To show that $d(t)$ is increasing in time, it suffices to show that $\alpha(t') + d(t)e$ separates from $\widetilde{\beta}(t')$ for $t' > t$. Let P be a point on the intersection of $\widetilde{\beta}(t)$ and $\alpha(t) + d(t)e$. We claim that, in a neighborhood of P, $\widetilde{\beta}^1(t')$ and $\alpha(t^1) + d(t)e$ are disjoint for $t' > t$. For, first let's assume P belongs to, interior of $\alpha(t) + d(t)e$. Then we can represent $\alpha(t')$ and $\widetilde{\beta}(t')$ locally at P on the graphs of two functions $x = x_1(y, t')$ and $x = x_2(y, t')$, respectively. By (6.2), both x_1 and

x_2 satisfy the equation

$$\frac{\partial x}{\partial t} = \left(1 + \left(\frac{\partial x}{\partial y}\right)^2\right)^{\frac{1}{2}}\left[\Phi(\theta)\frac{\partial^2 x/\partial y^2}{1 + (\partial x/\partial y)^2} + \Psi(\theta)\right]$$

in $[p_2 - \varepsilon, p_2 + \varepsilon] \times [t_1, t + \varepsilon]$ for some small $\varepsilon > 0$. Moreover, $x_2 \geqslant x_1 + d(t)$ at t and $x_2(p_2 \pm \varepsilon, t') > x_1(p_2 \pm \varepsilon, t')$ for $t' \in [t, t + \varepsilon]$. By the strong maximum principle,

$$x_2(y, t) > x_1(y, t) + d(t)$$

for (y, t') in $(p_2 - \varepsilon, p_2 + \varepsilon) \times (t, t + \varepsilon)$. So, $\alpha(t') + d(t)e$ is separated from $\widetilde{\beta}(t')$ in this neighborhood.

In case P happens to be one of the endpoints of $\alpha(t) + d(t)e$, we can represent the two arcs of $\gamma(\cdot, t)$ in a neighborhood of P, where one connects $\alpha(t')$ and the other connects $\widetilde{\beta}(t')$, as graphs of two functions $y = y_1(x, t')$ and $y = y_2(x, t')$, respectively. By the same reasoning as before, $\alpha(t^1) + d(t)e$ and $\widetilde{\beta}(t')$ separate within this neighborhood instantly.

Now, suppose that $\alpha(t') + d(t)e$ and $\widetilde{\beta}(t')$ do not separate instantly. We can find $\{t_j\}$, $t_j \downarrow t$, and $\{P_j\}$ on $\alpha(t^1) + d(t)e$ and $\widetilde{\beta}(t^1)$ such that $\{P_j\}$ converges to some P_0 on $\alpha(t) + d(t)e$. However, it means that $\alpha(t_j) + d(t)e$ is not separated from $\widetilde{\beta}(t_j)$ in any neighborhood of P_0 for all large j. This is contradictory to what we have proved. Thus, $\alpha(t') + d(t)e$ must separate from $\widetilde{\beta}(t')$ for all $t^1 > t$. $\square$

As an application of the whisker lemma, we have:

Proposition 6.9 *The total absolute curvature on a negative arc $c_j(t)$ tends to zero as $t \uparrow \omega$. Consequently, the total absolute curvature of $\gamma(\cdot, t)$ tend to 2π as $t \uparrow \omega$.*

Proof: Were there a negative arc whose total absolute curvature is greater than π in $[0, \omega)$, we can find a family of nice arcs $\alpha(t)$ on this arc in a suitably fixed coordinate system. But the existence of a δ-whisker prevents the curve from shrinking to a point. Hence, the total curvature along a negative arc must be less than $\pi - \delta$. By Lemma 6.7, its curvature is uniformly bounded. Since the total length of $\gamma(\cdot, t)$ tends to zero, we have

$$\int_{c_j} |k| ds \leqslant \text{ const. } \times L(t) \longrightarrow 0$$

as $t \uparrow \omega$.   □

6.5   The convexity theorem

Finally, we finish the proof of Theorem 6.1 in this section. Assume that the flow resists convexity until its very end. By a proper choice of coordinates we may assume, in view of Proposition 6.9, that, on every $\gamma(\cdot, t)$, there is an arc of negative curvature with normal image always containing the north pole and shrinking to it as $t \uparrow \omega$.

Let $\widetilde{\gamma}$ be an area preserving expansion of γ. We shall consider two cases separately.

Case I   There exists $\{t_j\}, t_j \uparrow \omega$, such that the diameter of each $\widetilde{\gamma}(\cdot, t_j)$ is uniformly bounded.

Case II   The diameter of $\widetilde{\gamma}(\cdot, t)$ tends to infinity as $t \uparrow \omega$.

As the total absolute curvature of $\gamma(\cdot, t)$ approaches 2π at the end, we may assume that $\gamma(\cdot, t)$ is the union of the graphs of two functions: $u_1(x, t) \leqslant u_2(x, t)$ for $x \in [a(t), b(t)]$. All arcs of negative curvature will eventually lie parallel to the x-axis, for otherwise, there would be an arc of positive curvature whose total curvature is strictly less than π, and yet has a positive lower bound. However, this is impossible by Lemma 6.7 and Theorem 6.6.

We shall first show that Case I is impossible.

For each $\gamma(\cdot, t_j)$, we can fix a coordinate system such that the origin is located on an arc of negative curvature and its unit tangent is $\boldsymbol{e}_1 = (1,0)$, and the normal $\boldsymbol{n} = (0,1)$ points to the interior of $\tilde{\gamma}(\cdot, t_j)$. In Case I, there exists $M > 0$ such that $\tilde{\gamma}(\cdot, t_j)$ is bounded between $x = \pm M$ and, $\tilde{u}_1(x, t_j) > -\varepsilon_j$ with $\varepsilon_j \longrightarrow 0$ as $j \longrightarrow \infty$. Here $\tilde{u}_1$ corresponds to u_1.

For each t_j, we change the time scale of $\tilde{\gamma}$ by setting

$$\bar{\gamma}(\cdot, \tau) = \sqrt{\frac{A(0)}{A(t_j)}} \gamma(\cdot, t) \,,$$

where $A(t)$ is the area enclosed by $\gamma(\cdot, t)$ and $\tau = A(0)(t - t_j)/A(t_j)$. Then, $\bar{\gamma}$ satisfies

$$\begin{cases} \dfrac{\partial \bar{\gamma}}{\partial t} = \left(\Phi \bar{k} + \mu \Psi \right) \boldsymbol{n} \,, & \mu = \sqrt{\dfrac{A(t_j)}{A(0)}} \,, \\ \bar{\gamma}(\cdot, 0) = \tilde{\gamma}(t_j) & \end{cases} \qquad (6.11)$$

for τ in $\left[0, A(0)(\omega - t_j)/A(t_j) \right)$.

By (1.19),

$$A(\omega) - A(t_j) = -\left(\int_0^{2\pi} \Phi(\theta) d\theta \right)(\omega - t_j) - \int_{t_j}^{\omega} \int_{\gamma} \Psi ds dt \,.$$

Therefore,

$$\lim_{t_j \uparrow \omega} \frac{A(t_j)}{\omega - t_j} = \int_0^{2\pi} \Phi(\theta) d\theta \,.$$

For all sufficiently large j, we may assume that (6.11) is valid in $[0, a]$ where

$$a = A(0) \left(2 \int \Phi(\theta) d\theta \right)^{-1} \,.$$

Let h be the solution for the mixed problem

$$\begin{cases} \dfrac{\partial h}{\partial t} = \dfrac{1}{4}\Phi_{min}\dfrac{\partial^2 h}{\partial x^2}\,, & (x,\tau) \in (-M-2,-M+2)\times[0,1)\,, \\[2ex] h(x,0) = f(x)\,, & x \in [-M-2, M+2]\,, \\[2ex] h(\pm(M+2),\tau) = 1\,, & \tau \in [0,1)\,, \end{cases}$$

$$(6.12)$$

where f is given by

$$f(x) = \begin{cases} \dfrac{1}{2}(x-M-2)+1\,, & x \in [M, M+2]\,, \\[2ex] 0 & x \in (-M, M)\,, \\[2ex] -\dfrac{1}{2}(x+M+2)+1\,, & x \in [-M-2, -M]\,. \end{cases}$$

Lemma 6.10 *For each $\delta \in (0,1)$, there exists $\rho > 0$ such that the solution of (6.12) satisfies*

(i) $h(x,\tau) \geqslant \rho$ for all $(x,\tau) \in [-M-2, M+2]\times[\delta,1)$, and

(ii) $|h_x(x,\tau)| \leqslant 1$ for all $(x,\tau) \in [-M-1, M+1]\times[0,1)$.

Proof: (i) is a direct consequence of the strong maximum principle. To prove (ii), we first observe that $h(\cdot, t)$ is convex for all t. Therefore, for all $x \in [-M-1, M+1]$,

$$|h_x(x,\tau)| \leqslant \frac{1-h(x,\tau)}{\text{dist}\{x, \pm(M+2)\}}$$

$$\leqslant 1\,. \qquad \square$$

Fix $\delta = \min\{1, a\}$ and $\varepsilon < \rho/2$. We choose a smooth approximation of f, g, satisfying $g'' > 0$, $f \leqslant g \leqslant f + \varepsilon$, and $g(\pm(M + 2)) = 1$.
Denote the solution of the mixed problem of (6.12) where f is re-
placed by g by $\widetilde{h}$. Then, $\widetilde{h}$ satisfies

$$\widetilde{h}(x, \tau) \geqslant \rho \text{ for all } (x, \tau) \in [-M - 2, M + 2] \times [\delta, 1)$$

and

$$\left|\widetilde{h}_x(x, \tau)\right| \leqslant 1 \text{ for all } (x, \tau) \in [-M - 1, M + 1] \times [0, 1)$$

for the same δ and ρ. Moreover, by applying the strong maximum
principle to the second derivatives of $\widetilde{h}$, we know that

$$\widetilde{k}(x, \tau) \geqslant k_0 > 0 , \quad \text{for all } (x, \tau) \in [-M - 1, M + 1] \times (0, 1) ,$$

for some k_0, where $\widetilde{k}(\cdot, \tau)$ is the curvature of the curve $(x, \widetilde{h}(x, \tau))$.
Setting $w = \widetilde{h} - 2\varepsilon$, by (6.12) we have

$$\frac{\partial w}{\partial \tau} = \frac{1}{4} \Phi_{\min} \frac{\partial^2 w}{\partial x^2}$$

$$\leqslant \frac{1}{2} \Phi_{\min} \frac{w_{xx}}{1 + w_x^2} .$$

By the construction of $\widetilde{h}$, we can find $\mu_0 > 0$ such that

$$\frac{\partial w}{\partial \tau} \leqslant \Phi \cdot \frac{w_{xx}}{1 + w_x^2} + \mu \Psi (1 + w_x^2)^{\frac{1}{2}}$$

in $[-M - 1, M + 1] \times [0, 1)$ for all $\mu \in [0, \mu_0]$.

For large j, $\widetilde{\gamma}(\cdot, t)$ is bounded between $x = \pm M$ and lies above
the graph of $g - 2\varepsilon$. Let's consider the flow $\overline{\gamma}(\cdot, \tau)$ and look at the
arc containing the origin. It is the graph of some function $\overline{u}(x, \tau)$.

Since the curvature is negative there, by (6.3) and (1.3) we have

$$\frac{\partial \bar{u}}{\partial \tau} \leqslant \mu \Psi \left(1 + \left(\frac{\partial \bar{u}}{\partial x}\right)^2\right)^{1/2}$$

$$\leqslant \mu C |\Psi|_{\max}$$

as the curvature is always bounded on this arc. Therefore, at time τ, this subarc lies below the line $y = \mu C |\Psi|_{\max}\tau$. On the other hand, by applying the comparison principle to $\bar{u}$ and w, we know that this arc must lie above the graph of w at any τ. Taking, in particular, $\tau = \delta$, we have

$$\bar{u}(x,\delta) \geqslant \tilde{h}(x,\delta) - 2\varepsilon$$

$$\geqslant \rho - 2\varepsilon .$$

Hence,

$$\rho - 2\varepsilon \leqslant \mu c |\Psi|_{\max}\delta .$$

However, the right-hand side of this inequality tends to zero as $j \longrightarrow \infty$, and the contradiction holds. So Case I is excluded.

Next, we consider Case II.

The difference $U = u_2 - u_1$ satisfies

$$\frac{\partial U}{\partial t} = A(x,t)\frac{\partial^2 U}{\partial x^2} + B(x,t)\frac{\partial U}{\partial x} , \tag{6.13}$$

for some $A(x,t) > 0$ and $B(x,t)$, $x \in [a(t), b(t)]$.

Lemma 6.11 *There exists t_0 close to ω such that $U(\cdot,t)$ has only one local maximum in $[a(t), b(t)]$.*

Proof: We shall prove the lemma by showing that U has no local minima in $(a(t), b(t))$ for all $t \geqslant t_0$ where t_0 is close to ω. By applying the Sturm oscillation theorem to the equation satisfied by $\partial U/\partial x$, we know that the number of local minima is finite and nonincreasing in t. Moreover, through every minimum P there passes a uniquely determined differentiable curve $X(t)$ consisting of local minima of $\gamma(\cdot, t)$. Were the lemma not true, there exists such path $X(t)$ for all $t \in [t_0, \omega)$. By (6.13),

$$\frac{dU}{dt}\big(X(t), t\big) = \frac{\partial U}{\partial t}\big(X(t), t\big) + \frac{\partial U}{\partial x}\big(X(t), t\big)\frac{dX}{dt}(t) \geqslant 0 \ .$$

Hence,

$$0 < U\big(X(t_0), t_0\big) \leqslant U\big(X(t), t\big) \longrightarrow 0$$

as $t \uparrow \omega$, which is impossible. $\qquad \square$

Lemma 6.12 *Let γ be any embedded closed curve which is the union of the graphs of two functions u_1 and u_2, $u_1 \leqslant u_2$, over $[a, b]$. Suppose that $U \equiv u_2 - u_1$ has only one local maximum. Then,*

$$q_2 - q_1 \leqslant \frac{2}{3}(b - a) \ ,$$

where q_1 and q_2, $q_1 < q_2$, are respectively the x-coordinates of the vertical lines which separate γ into three pieces, the left and the right ones having one quarter of the area.

Proof: Let $\bar{q}$ be the x-coordinate of the vertical line which divides γ, or, more precisely, the region enclosed by γ, into two pieces of equal area. The vertical lines through $q_1, \bar{q}$, and q_2 divide the region into four parts of equal area. Call them P_1, P_2, P_3, and P_4, and denote their widths, that is, the differences between the q's and a, b, by w_1, w_2, w_3, and w_4. The maximum thickness, $U_{\max} = \max\{u_2(x) - u_1(x) : x \in [a, b]\}$, of γ is realized either in P_1, P_4 or P_2, P_3.

Suppose that $U_{\max}$ is realized in P_1. By Lemma 6.11, U decreases as we move through the other Ps. As all P_is have equal area, $w_2 \leqslant w_3 \leqslant w_4$. Therefore,

$$
\begin{aligned}
q_2 - q_1 &= w_2 + w_3 \\
&\leqslant \frac{2}{3}(w_2 + w_3 + w_4) \\
&\leqslant \frac{2}{3}(b - a) \ .
\end{aligned}
$$

Suppose that $U_{\max}$ is realized in P_2. We may assume the area of P_i, $i = 1, \cdots, 4$, is equal to 1. Divide P_2 further into two parts by a vertical line through the maximal thickness. Denote their areas and widths by A_1, A_3 and w_{21}, w_{23}, respectively. Let U_1 and U_3 be the values of $U(x)$, where x denotes the left and the right endpoints of P_2, respectively. By Lemma 6.11,

$$
U \leqslant U_1 \text{ on } w_1 \ , \qquad U \geqslant U_1 \text{ on } w_{21} \ ,
$$

$$
U \geqslant U_3 \text{ on } w_{23} \ , \qquad U \leqslant U_3 \text{ on } w_3 \ .
$$

Therefore,

$$
w_1 \geqslant \frac{1}{U_1} \ , \quad w_{21} \leqslant \frac{A_1}{U_1}
$$

and

$$
w_{23} \leqslant \frac{A_3}{U_3} \ , \quad w_3 \geqslant \frac{1}{U_3} \ .
$$

It follows that

$$
w_{21} \leqslant A_1 w_1 \leqslant w_1 \ ,
$$

and

$$
w_{23} \leqslant A_3 w_3 \leqslant w_3 \ .
$$

As before, $w_3 \leqslant w_4$, and so,

$$
\begin{aligned}
q_2 - q_1 &= w_{21} + w_{23} + w_3 \\
&\leqslant w_1 + w_2 + w_3 \\
&< 2w_1 + 2w_4 \, ,
\end{aligned}
$$

which implies that

$$
3(q_2 - q_1) < 2(w_1 + w_2 + w_3 + w_4) \, .
$$

□

In Case II, the diameter of $\tilde{\gamma}(\cdot, t)$ tends to infinity as $t \uparrow \omega$. For each $\tilde{\gamma}(\cdot, t)$, there exists a translation so that the origin is located on an arc of negative curvature with tangent $\mathbf{e}_1$. When there exist M, β, and $\{t_j\}, t_j \uparrow \omega$, such that each $\tilde{\gamma}(\cdot, t_j)$, after a translation, lies between the lines $y = \beta(x - M)$, $y = -\beta(x + M)$, and $\tilde{u}_1 > -\varepsilon_j$, with $\varepsilon_j \longrightarrow 0$ as $j \longrightarrow \infty$, we can follow the argument in Case I to draw a contradiction. Here, $\tilde{u}_1$ corresponds to the dilation of u_1. Therefore, we may assume, for all large M and small $\beta > 0$, each $\tilde{\gamma}(\cdot, t)$, after a suitable translation, touches one of the lines $y = \beta(\pm x - M)$. Let $\hat{\gamma}(\cdot, t)$ be the length-preserving expansion of $\gamma(\cdot, t)$. By noting the facts that $\hat{\gamma}(\cdot, t)$ has fixed enclosed area and its total absolute curvature tends to 2π, we know that $\hat{\gamma}(\cdot, t)$ must collapse into a line segment on the x-axis after a suitable translation depending on t.

Let ℓ_1 and ℓ_2 be the vertical lines $x = a(t)$ and $x = b(t)$, respectively. Any vertical line $x = \ell$ between ℓ_1 and ℓ_2 cuts u_1 (resp. u_2) at one point transversely. Denote its slope by $\tan \alpha_i$, $|\alpha_i| < \pi/2$, $i = 1, 2$. We claim that, for each fixed $\varepsilon > 0$, there exists t'' such that, if $|\alpha_i| > \varepsilon$ for some i, the minimum distance between ℓ and ℓ_1, and the distance between ℓ and ℓ_2, is less than $\varepsilon(b(t) - a(t))$ for all t in $[t'', \omega)$. For, we can look at the length-preserving expansion $\hat{\gamma}(\cdot, t)$ which collapses into a line segment after a time-dependent translation. Since its total absolute curvature tends to 2π, the slopes at the intersections of any vertical lines based at a point $[\widehat{a}(t) + \varepsilon, \widehat{b}(t) - \varepsilon]$ and $\widehat{\gamma}(\cdot, t)$ must tend to zero as $t \uparrow \omega$. Hence, there exists t'' such that $|\alpha_i| \leqslant |\tan \alpha_i| < \varepsilon$.

Now, for some small ε to be specified below, we consider $t \geqslant \max\{t_0, t''\}$, where t_0 is chosen according to Lemma 6.11. Let

$$\widehat{t} = t + \frac{\left(\frac{1}{2} + \frac{\Phi_{\max}}{\Phi_{\min}}\varepsilon\right) A(t)}{\int \Phi(\theta)d\theta} .$$

We have

$$
\begin{aligned}
A(\widehat{t}) &= A(t) - \int_t^{\widehat{t}} \int_\gamma \left(\Phi k + \Psi\right) ds d\tau \\
&= A(t) - \left(\int \Phi(\theta)d\theta\right)\left(\frac{1}{2} + \frac{\Phi_{\max}}{\Phi_{\min}}\varepsilon\right)\frac{A(t)}{\int \Phi(\theta)d\theta} \\
&\quad - \int_t^{\widehat{t}} \int_\gamma \Psi ds dt \\
&\geqslant \left[\frac{1}{2} - \frac{\Phi_{\max}}{\Phi_{\min}}\varepsilon - \left(\frac{1}{2} + \frac{\Phi_{\max}}{\Phi_{\min}}\varepsilon\right)\frac{|\Psi|_{\max}}{\int \Phi d\theta}\max_{[t,\widehat{t}]} L(\tau)\right] A(t) ,
\end{aligned}
$$

where $L(t)$ is the length of $\gamma(\cdot, t)$.

As $L(t)$ tends to zero as $t \uparrow \omega$, for small ε and t close to ω,

$$A(\widehat{t}) \geqslant \frac{5}{11}A(t) . \qquad (6.14)$$

Next, for a fixed t, we consider q_1 and q_2 for $\gamma(\cdot, t)$ as determined in Lemma 6.12. The vertical lines passing q_1 and q_2 divide $\gamma(\cdot, t)$, $\tau \in [t, \widehat{t}]$, into, at most, three parts. At the left piece, either

(i) for all $\tau \in [t, \widehat{t}]$, the angles α made with the tangents at the intersection of $x = q_1$ and $\gamma(\cdot, \tau)$ do not exceed ε, or

(ii) there exists $\tau \in [t, \hat{t}]$ such that one of these angles satisfies $|\alpha| > \varepsilon$.

In the first case, denote the area of this piece by $A_\ell(\tau)$. We have

$$-\frac{dA_\ell(\tau)}{d\tau} = \int_{\pi-\alpha}^{2\pi+\alpha} \Phi(\theta)d\theta - \int_{C(\cdot,\tau)} \Psi ds$$

$$\geqslant \int_{\pi+\varepsilon}^{2\pi-\varepsilon} \Phi(\theta)d\theta - |\Psi|_{\max} \cdot L(\tau) ,$$

where $C(\cdot, \tau)$ is the arc $\gamma(\cdot, \tau) \cap \{(x, y) : x \leqslant q_1\}$. We have

$$A_\ell(\hat{t})$$

$$\leqslant A_\ell(t) - \left(\int_{\pi+\varepsilon}^{2\pi-\varepsilon} \Phi d\theta - |\Psi|_{\max} \max L(\tau) \right) \left(\frac{1}{2} + \frac{\Phi_{\max}}{\Phi_{\min}} \varepsilon \right) \frac{A(t)}{\int \Phi d\theta}$$

$$= \frac{A(t)}{\int \Phi d\theta} \left[\frac{1}{4} \int \Phi d\theta - \left(\int_{\pi+\varepsilon}^{2\pi-\varepsilon} \Phi d\theta - |\Psi|_{\max} \max L(\tau) \right) \left(\frac{1}{2} + \frac{\Phi_{\max}}{\Phi_{\min}} \varepsilon \right) \right]$$

$$\leqslant \frac{A(t)}{\int \Phi d\theta} \left[(1 + 2\varepsilon - \pi)\Phi_{\max}\varepsilon + \frac{1}{2}|\Psi|_{\max} \max L(\tau) + \right.$$

$$\left. \varepsilon |\Psi|_{\max} \frac{\Phi_{\max}}{\Phi_{\min}} \max L(\tau) \right] .$$

Notice that we have used (6.2) to get

$$\int_\pi^{2\pi} \Phi = \frac{1}{2} \int_0^{2\pi} \Phi .$$

After a further restriction on t, we find that, for small ε,

$$A_\ell(\hat{t}) < 0 ,$$

which is impossible.

So, with this choice of ε, Case (ii) must hold. In other words, there is some $\tau \in [t, \hat{t}]$ such that

$$q_1 - a(\tau) \leqslant \varepsilon\big(b(\tau) - a(\tau)\big)$$
$$\equiv \varepsilon w(\tau) .$$

Clearly, under (6.3), straight lines translate in constant speed. By comparing $\gamma(\cdot, t)$ with the evolution of the vertical line at $b(t)$, we have, for $t_1 < t_2$,

$$a(t_2) \geqslant a(t_1) - |\Psi|_{\max}(t_2 - t_1) ,$$

$$b(t_2) \leqslant b(t_1) + |\Psi|_{\max}(t_2 - t_1) .$$

Therefore,

$$\begin{aligned} q_1 - a(\hat{t}) &\leqslant q_1 - a(\tau) + |\Psi|_{\max}(\hat{t} - \tau) \\ &\leqslant \varepsilon w(\tau) + |\Psi|_{\max}(\hat{t} - \tau) \\ &\leqslant \varepsilon w(t) + 2\varepsilon|\Psi|_{\max}(\tau - t) + |\Psi|_{\max}(\hat{t} - \tau) \\ &\leqslant \varepsilon w(t) + (1 + 2\varepsilon)|\Psi|_{\max}(\hat{t} - \tau) . \end{aligned}$$

Similarly, we have

$$b(\hat{t}) - q_2 \leqslant \varepsilon w(t) + (1 + 2\varepsilon)|\Psi|_{\max}(\hat{t} - \tau) .$$

It follows from the isoperimetric inequality $Cw^2(t) \geqslant A(t)$ that we have, by Lemma 6.12,

$$
w(\hat{t}) \;\leqslant\; q_2 - q_1 + 2\varepsilon w(t) + 2(1 + 2\varepsilon)|\Psi|_{\max}(\hat{t} - \tau)
$$

$$
= \; q_2 - q_1 + 2\varepsilon w(t) + 2(1 + 2\varepsilon)|\Psi|_{\max}\left(\frac{1}{2} + \frac{\Phi_{\max}}{\Phi_{\min}}\varepsilon\right) \times \frac{A(t)}{\displaystyle\int \Phi d\theta}
$$

$$
\leqslant \; \left(\frac{2}{3} + 2\varepsilon + \widetilde{C}w(t)\right)w(t) \; ,
$$

where $\widetilde{C}$ is a positive constant depending only on Φ and Ψ. As $w(t)$ tends to zero, for every sufficiently small ε, there exists t^* sufficiently close to ω such that

$$
\frac{2}{3} + 2\varepsilon + \widetilde{c}w(t) \leqslant \frac{2}{3} \times \frac{100}{99} \; .
$$

Therefore,

$$
w(\hat{t}) \leqslant \frac{200}{297}w(t) \; .
$$

Taking $t_1 = t^*$,

$$
t_{j+1} = t_j + \left(\frac{1}{2} + \frac{\Phi_{\max}}{\Phi_{\min}}\varepsilon\right)\frac{A(t)}{\displaystyle\int \Phi d\theta} \; , \quad \text{for } j \geqslant 1 \; ,
$$

we have

$$
\frac{A(t_{j+1})}{w^2(t_{j+1})} \;\geqslant\; \frac{\frac{5}{11}A(t_{j+1})}{\left(\frac{200}{297}\right)^2 w^2(t_{j+1})}
$$

$$
> \; 1.002\frac{A(t_j)}{w^2(t_j)} \; .
$$

So,

$$
\frac{A(t_{j+1})}{w^2(t_{j+1})} \geqslant (1.002)^j \frac{A(t^*)}{w^2(t^*)} \longrightarrow \infty \quad \text{as} \quad j \longrightarrow \infty \; ,
$$

and yet, on the other hand,

$$\frac{A(t_{j+1})}{w^2(t_{j+1})} \longrightarrow 0 \ , \quad \text{as} \ \ j \longrightarrow \infty \ ,$$

as $\widehat{\gamma}(\cdot, t)$ converges to a line segment. This contradiction finally shows that Theorem 6.1 must hold.

Notes

The content of this chapter is based on Grayson [66], Angenent [13], [14], Oaks [92], and Chou-Zhu [36]. In particular, Theorem 6.4 is taken from [13], Theorem 6.6 from [92], and Theorem 6.1 from [36].

The limit curve γ^*. Theorem 6.4 continues to hold for immersed closed flows of (1.21) on a surface under assumptions (i)–(vi) (see Notes in Chapter 1). The behaviour of this flow near the singularities is studied in some depth in [14] and [92]. Let's follow the latter and describe a basic result. First, we call $[a, b]$ a *singular interval* of the flow if it is the largest interval satisfying the following two conditions: (a) $k([a, b], t)$ becomes unbounded as $t \uparrow w$ and (b) $\gamma^*([a, b])$ is a singularity. Let $Q = \gamma^*(p_0)$. For any $\varepsilon > 0$, we denote by $[a_\varepsilon, b_\varepsilon]$ the maximal arc containing p_0 over which γ converges entirely inside $N_\varepsilon(Q)$. We have

Theorem *Let $\gamma(\cdot, t)$ be a maximal flow of (1.21) where (i)–(vi) hold and w is finite. Then,*

(A) if $\gamma(\cdot, t)$ is embedded, then it shrinks to a point as $t \uparrow w$,

(B) let $Q = \gamma^*(p_0)$ be a singularity. Then, p_0 is contained in some singular interval I and either (a) there is a self-intersection in $\gamma([a_\varepsilon, b_\varepsilon], t)$ converging to Q ("a loop contracts"), or (b) there are $p_1, p_2 \notin I$ such that $\gamma^*([p_1, p_0]) = \gamma^*([p_0, p_2])$ ("parametrization doubling").

It is commonly believed that subcase (B)(b) cannot happen.

Convexity result. Theorem 6.1 was first conjectured in Gage [57] when $\Psi \equiv 0$. After proving that the flow shrinks to a point, the proof of convexity follows Grayson [66] closely. Nevertheless, one can show that Hamilton's approach (see Chapter 5) can be extended to the anisotropic case. In fact, Zhu [113] has used this approach to establish the convexity and asymptotic behaviour of the flow (6.3) on a surface.

Chapter 7

Embedded Closed Geodesics on Surfaces

As an application to the theory of curve shortening flows developed in the previous chapters, we shall prove the existence of embedded closed geodesics on a closed surface.

A standard approach for finding closed geodesics is to look for curves which minimize the length among a given homotopy class. However, for a simply-connected surface, the homotopy class is trivial and this approach fails. When the surface is convex, Poincaré proposed to look for a geodesic by minimizing length among all embedded closed curves which divide the surface into two pieces each having total Gaussian curvature 2π. Gage [56] found a certain curve shortening flow which preserves curves in this class. It turns out that this flow does exist for all time and subconverges to an embedded closed geodesic. For a general closed surface, the same proof shows that the standard curve shortening flow either exists for all time or shrinks to a point in finite time. Together with a topological minimax argument, one can show that there are always three embedded closed geodesics on a 2-sphere—a result first formulated by Lusternik and Schnirelmann in 1929.

7.1   Basic results

Let M be an oriented smooth surface. Here, surface always means a two-dimensional Riemannian manifold with the metric g. Consider a family of smooth closed curves $\gamma = \gamma(u,t) : S^1 \times [0,T] \longrightarrow M$. Denote by $\partial\gamma/\partial t = \gamma_*(\partial/\partial t)$ its velocity vector field and $\partial\gamma/\partial u = \gamma_*(\partial/\partial u)$ its tangent vector field. The unit tangent vector $\boldsymbol{T}$ is given by $\gamma_u/|\gamma_u|$ and the unit normal vector $\boldsymbol{N}$ is chosen such that $(\boldsymbol{T}, \boldsymbol{N})$ agrees with the orientation of the surface.

Given a smooth function F defined in $M \times \mathbb{R}$, consider the Cauchy problem

$$\frac{\partial\gamma}{\partial t} = F(\gamma, k)\boldsymbol{N} \ , \quad \gamma_0 \text{ embedded closed }, \qquad (7.1)$$

where k is the geodesic curvature of the curve γ. In the following, we let $s = |\gamma_u|$.

Lemma 7.1 *Let $\gamma(\cdot, t)$ be a solution of (7.1). Then, the following hold:*

(i) $\dfrac{\partial s}{\partial t} = -Fks,$

(ii) $\dfrac{\partial \boldsymbol{T}}{\partial t} = F_s \boldsymbol{N},$

(iii) $\left[\dfrac{\partial}{\partial t}, \dfrac{\partial}{\partial s}\right] = kF\dfrac{\partial}{\partial s},$

(iv) $\dfrac{\partial k}{\partial t} = F_{ss} + k^2 F + KF,$

(v) $\dfrac{d}{dt}\displaystyle\int_{\gamma(\cdot,t)} k\, ds = \int_\gamma KF\, ds,$

(vi) $\dfrac{d}{dt} \displaystyle\int\limits_{\gamma(\cdot,t)} ds = - \displaystyle\int\limits_{\gamma(\cdot,t)} Fk\,ds,$

where $s = s(\cdot, t)$ is the arc-length parametrization of $\gamma(\cdot, t)$ and K is the Gaussian curvature of M.

Proof: Recall that $\partial/\partial s = s^{-1}\partial/\partial u$. Let

$$\frac{\partial^2 \gamma}{\partial t \partial u} = \nabla_Z\left(\gamma_*\left(\frac{\partial}{\partial u}\right)\right),$$

$$\frac{\partial^2 \gamma}{\partial u \partial t} = \nabla_{\gamma_u}\left(\gamma_*\left(\frac{\partial}{\partial t}\right)\right),$$

$$Z = \gamma_*\left(\frac{\partial}{\partial t}\right),$$

where ∇ is the Levi-Civita connection on M. From $\gamma_{tu} = \gamma_{ut}$ and the Frenet formulas (1.1), which hold on surfaces, we have (i) and (ii).

Next,

$$\left[\frac{\partial}{\partial t}, \frac{\partial}{\partial s}\right]f = \frac{\partial}{\partial t}\left(\frac{1}{s}\frac{\partial f}{\partial u}\right) - \frac{1}{s}\frac{\partial}{\partial u}\left(\frac{\partial f}{\partial t}\right)$$

$$= kF\frac{\partial f}{\partial s},$$

by (i). Hence, (iii) follows.

In addition, by using the definition of the Riemannian curvature tensor and (iii), we have

$$\frac{\partial}{\partial t}(k\boldsymbol{N}) = \nabla_Z \nabla_{\boldsymbol{T}} \boldsymbol{T}$$

$$= \nabla_{\boldsymbol{T}} \nabla_Z \boldsymbol{T} + \nabla_{[Z,\boldsymbol{T}]}\boldsymbol{T} + R(Z, \boldsymbol{T})\boldsymbol{T}$$

$$= \nabla_{\boldsymbol{T}}(F_s \boldsymbol{N}) + \nabla_{[\partial/\partial t, \partial/\partial s]} \boldsymbol{T} + R(Z, \boldsymbol{T})\boldsymbol{T}$$

$$= F_{ss}\boldsymbol{T} - F_s k \boldsymbol{T} + F k^2 \boldsymbol{N} + R(Z, \boldsymbol{T})\boldsymbol{T} \ .$$

On the other hand, by (7.1),

$$\frac{\partial}{\partial t}(k\boldsymbol{N}) = k_t \boldsymbol{N} + k \frac{\partial \boldsymbol{N}}{\partial t} \ .$$

Taking the inner product with $\boldsymbol{N}$ yields (iv). Finally, (v) and (vi) follow from (i) and (iv). $\square$

Now we show that the solution of (7.1) exists for some positive time. Let the initial curve γ_0 be smooth and closed. We extend its parameterization to an immersion $\sigma : S^1 \times [-1, 1] \longrightarrow M$ with $\sigma\big|_{S^1 \times \{0\}} = \gamma_0$. Then, any smooth closed curve which is C^2-close to γ_0 can be parametrized as $\gamma_f(u) = \sigma(u, f(u))$ for some C^2-function f with $|f(u)| < 1$.

Consider a smooth solution $\gamma(\cdot, t) : S^1 \times [0, T) \longrightarrow M$ of (7.1) starting at γ_0. For small t, the solution can be represented as the image under σ of the graph of a function $f(u, t)$, i.e., $\gamma(u, t) = \sigma(u, f(u, t))$. The pull-back of the metric g on M under σ is given by

$$\sigma^*(g) = A(u, v)(du)^2 + 2B(u, v)dudv + C(u, v)(dv)^2 \ ,$$

where A, B, and C are smooth in $S^1 \times [-1, 1]$, and $D = AC - B^2 > 0$. Setting $\ell = A + 2Bf_u + Cf_u^2$, the tangent $\boldsymbol{T}$ and normal $\boldsymbol{N}$ to $\gamma(\cdot, t)$ are given, respectively, by

$$\boldsymbol{T} = d\sigma\Big(\ell^{-\frac{1}{2}}\big(\frac{\partial}{\partial u} + f_u \frac{\partial}{\partial v}\big)\Big) \quad \text{and}$$

$$\boldsymbol{N} = d\sigma\Big\{(\ell D)^{-\frac{1}{2}}\big[-(B + Cf_u)\frac{\partial}{\partial u} + (A + Bf_u)\frac{\partial}{\partial v}\big]\Big\} \ ,$$

where A, B, C, and D are evaluated at $(u, f(u,t))$. By the Frenet formulas, we have

$$k = \ell^{-\frac{3}{2}} D^{\frac{1}{2}} \left(f_{uu} + P + Q f_u + R f_u^2 + S f_u^3 \right) , \tag{7.2}$$

where P, Q, R, and S are smooth functions evaluated at $(u, f(u,t))$.

The vertical velocity of $\gamma(\cdot, t)$ is given by $f_t \partial/\partial v$. So, its normal velocity is

$$\sigma^* g\left(f_t \frac{\partial}{\partial v} , (\ell D)^{-\frac{1}{2}} \left[-(B + C f_u) \frac{\partial}{\partial u} + (A + B f_u) \frac{\partial}{\partial v} \right] \right)$$

$$= \ell^{-\frac{1}{2}} D^{\frac{1}{2}} f_t .$$

It follows from these computations that $\gamma(\cdot, t)$ solves (7.1) if and only if f solves

$$f_t = \ell^{\frac{1}{2}} D^{-\frac{1}{2}} F(\sigma, k) . \tag{7.3}$$

where $\ell, D, \cdots$ etc. are evaluated at $f(u,t)$.

¿From now on, we shall assume F in (7.1) is of the form

$$F = ak , \tag{7.4}$$

where a is a smooth, positive function in M. Then, equation (7.3) is a quasilinear parabolic equation under this assumption. It follows from Fact 3 in §1.2 and the proof of Proposition 1.2 that the following result holds.

Proposition 7.2 *Consider the Cauchy problem for (7.1) where F satisfies (7.4) and γ_0 is a smooth, closed curve. Then, it admits a unique maximal solution $\gamma(\cdot, t)$ in $M \times [0, \omega)$. Moreover, when ω is finite, the geodesic curvature of $\gamma(\cdot, t)$ becomes unbounded as $t \uparrow \omega$.*

The next crucial lemma shows that the total absolute of the solution remains bounded on any finite time interval.

Lemma 7.3 *We have*

$$\frac{d}{dt}\left(e^{-\mu t}\int_{\gamma(\cdot,t)}|k|ds\right) \leqslant -2e^{-\mu t}\sum_{j}\left|F_s(u_j(t),t)\right| ,$$

where the summation in the right-hand side of this inequality is over all inflection points $u_j(t)$ of $\gamma(\cdot,t)$ and $\mu = (aK)_{\max}$.

Notice that the Sturm oscillation theorem can be applied to the equation in Lemma 7.1 (iv) to show that the number of inflection points on $\gamma(\cdot,t)$ is finite for all $t \in (0,\omega)$.

Proof: By Lemma 7.1,

$$\frac{d}{dt}\int_{\gamma(\cdot,t)}|k|ds \;=\; -2\sum\left|F_s(u_j(t),t)\right| + \int_{k\geqslant 0} aKk\,ds - \int_{k\leqslant 0} aKk\,ds$$

$$\leqslant \;-2\sum\left|F_s(u_j(t),t)\right| + (aK)_{\max}\int_{\gamma(\cdot,t)}|k|ds ,$$

and the lemma follows.   □

The following lemma is similar to Proposition 1.10.

Lemma 7.4 *There exist positive constants δ and C depending only on M and a such that, for any solution $\gamma(\cdot,t)$ of (7.1),*

$$\gamma(\cdot,t) \subseteq N_{C\sqrt{t}}\big(\gamma(\cdot,0)\big) ,$$

for $0 \leqslant t < \min\{\delta,\omega\}$.

Proof: Let P be any point in M which does not lie on the initial curve $\gamma(\cdot,0)$. Let $d(t)$ be the distance from P to $\gamma(\cdot,t)$. Since the solution is smooth, d is Lipschitz continuous in t.

Let $\rho_0 > 0$ be the injectivity radius of M. That means the exponential map $\exp_P : T_P M \longrightarrow M$ is an embedding on the disk D_{ρ_0} centered at the origin of $T_P M$. If, at some instant t, the distance function is less than $\rho_0/2$, we can choose Q on $\gamma(\cdot, t)$ which minimizes dist $(\gamma(\cdot, t), P)$. This point must lie in the image of D_{ρ_0} under the exponential map.

Let (r, φ) be the geodesic polar coordinates at P. We can represent the solution near Q as a graph $r = f(\varphi, t)$, where $Q = (d(t), \varphi_0)$. Then $f(\varphi_0, t) = d(t)$ and $\partial f/\partial\varphi(\varphi_0, t) = 0$. Moreover, $\partial^2 f/\partial\varphi^2(\varphi_0, t) \geqslant 0$. In these geodesic polar coordinates, the metric g is given by

$$g = (dr)^2 + A(r, \varphi)(d\varphi)^2 \ .$$

By a direct computation, the geodesic curvature of the graph of $r = f(\varphi, t)$ is

$$k = \frac{A^{\frac{1}{2}}}{(A + p^2)^{\frac{3}{2}}} \left(\frac{\partial^2 f}{\partial\varphi^2} - \frac{A_r}{A} \left(\frac{\partial f}{\partial\varphi} \right)^2 - \frac{A_\varphi}{2A} \frac{\partial f}{\partial\varphi} - \frac{1}{2} A_r \right),$$

where A_φ and A_r are evaluated at $r = f(\varphi, t)$. Hence, the geodesic curvature of $\gamma(\cdot, t)$ at Q is at least the curvature of the geodesic circle with radius $d(t)$ centered at P. On the other hand, it is not hard to see that the curvature of a geodesic circle with radius r, $r \leqslant \rho_0/2$, is not less than $-\beta^2/(2r)$ for some β depending only on the surface. Therefore, by (7.1), the distance function satisfies

$$d'(t) \geqslant -\frac{C_0}{d(t)} \ , \tag{7.5}$$

whenever $d(t) \leqslant \rho_0/2$. Here, C_0 depends on M and F only.

Now, let $\delta = \rho_0^2/(8C_0)$. If P lies on $\gamma(\cdot, t)$ for some t, then $d(t) = 0$. Integrating (7.5) gives

$$0 = d^2(t) \geqslant d^2(0) - 2C_0 t \ ,$$

which means that P lies in a $\sqrt{2C_0 t}$−neighborhood of $\gamma(\cdot, 0)$.   □

7.2   The limit curve

In this and the next section, we shall show that the flow (7.1) shrinks to a point when ω is finite. As usual, we shall assume this is not true and draw a contradiction by analyzing the flow near a singularity.

First of all, when γ_0 is embedded and closed, we note that $\gamma(\cdot, t)$ is also embedded. This is a consequence of Proposition 1.5. Suppose ω is finite. By the same argument as in the proof of Chapter 6, as $t \uparrow \omega$, the solution converges in the Hausdorff metric to a limit curve γ^* which is the image of a Lipschitz continuous map from S^1 to M. Moreover, after a slight modification of the proof of Theorem 6.1, we deduce from Lemma 7.3 that there exists a finite number of singularities $\{Q_1, \cdots, Q_m\}$ on γ^*, such that $\gamma^* \setminus \{Q_1, \cdots, Q_m\}$ consists of smooth curves. Away from the singularities, the curve $\gamma(\cdot, t)$ converges to γ^* smoothly. Furthermore, for any P in $\gamma^* \setminus \{Q_1, \cdots, Q_m\}$, there exists a neighborhood U of P and $t_0 \in [0, \omega)$ such that $U \cap \gamma(\cdot, t)$ is a connected evolving arc for all $t \geqslant t_0$. On the other hand, for any singularity Q, we can find t_1 close to ω and a small $\rho_1 > 0$ such that the evolving arc $\Gamma(\cdot, t) = D_{2\rho_1}(Q) \cap \gamma(\cdot, t)$, where $D_{2\rho_1}(Q)$ is the geodesic disk of radius $2\rho_1$ at Q, is connected, and has exactly two endpoints converging to two distinct points on $\gamma^* \cap \partial D_{2\rho_1}(Q)$ as $t \uparrow \omega$. We employ isothermal coordinates to express the evolution of $\Gamma(\cdot, t)$ on M as an evolution in the plane.

Without loss of generality, we may assume $D_{2\rho_1}(Q)$ is covered by isothermal coordinates. So, there is a conformal diffeomorphism ϕ from $D_{2\rho_1}(Q)$ to some open set $V \subseteq \mathbb{R}^2$ such that the metric g becomes

$$g = J^2(x, y)(dx^2 + dy^2) \, ,$$

where J is bounded between two positive constants. Let $X = J^{-1}\partial/\partial x$ and $Y = J^{-1}\partial/\partial y$ be the unit vectors. Let $\Gamma'(u,t) = \phi(\Gamma(u,t))$. Since ϕ is conformal, $\phi_*(\boldsymbol{N}) = J^{-1}\boldsymbol{n}'$, where $\boldsymbol{n}'$ is the unit normal of the plane curve $\Gamma'(\cdot,t)$. It implies that $\Gamma'(\cdot,t)$ evolves according to

$$\frac{\partial \Gamma'}{\partial t} = \frac{1}{J}F\boldsymbol{n}' . \tag{7.6}$$

It is well-known that the Christoffel symbols in isothermal coordinates are given by the following relations:

$$\nabla_X X = -\left(\frac{J_Y}{J}\right)Y \ , \quad \nabla_X Y = \left(\frac{J_Y}{J}\right)X \ ,$$

$$\nabla_Y X = \left(\frac{J_X}{J}\right)Y \quad , \quad \nabla_Y Y = -\left(\frac{J_X}{J}\right)X \ , \tag{7.7}$$

where $J_X = \nabla_X J$ and $J_Y = \nabla_Y J$. Let θ be the angle between $\boldsymbol{T}$ and X. So,

$$\boldsymbol{T} = \cos\theta X + \sin\theta Y$$

and

$$\boldsymbol{N} = -\sin\theta X + \cos\theta Y \ . \tag{7.8}$$

By (7.7) and (7.8), we have

$$k\boldsymbol{N} = \nabla_{\boldsymbol{T}}\boldsymbol{T}$$

$$= \left(\frac{1}{J}k' - \frac{J_Y}{J}\cos\theta + \frac{J_X}{J}\sin\theta\right)\boldsymbol{N} \ ,$$

where k' is the curvature of $\Gamma'(\cdot,t)$. Hence, the Liouville formula,

$$k = \frac{1}{J}k' - \frac{J_Y}{J}\cos\theta + \frac{J_X}{J}\sin\theta \ , \tag{7.9}$$

holds. In view of this, the flow (7.6) can be written in the form

$$\frac{\partial \Gamma'}{\partial t} = \left(\Phi k' + \Psi \right) n' \, . \tag{7.10}$$

where $\Phi = \Phi(x, y)$ is pinched between two positive constants and Ψ satisfies

$$\Psi(x, y, \theta + \pi) = -\Psi(x, y, \theta) \, , \quad \forall (x, y, \theta) \in V \times S^1 \, .$$

In other words, (7.10) is uniformly parabolic and symmetric.

Without loss of generality, we may assume that $\phi(Q)$ is the origin and V contains the unit disk centered at the origin D_1. Also, the intersection of $\Gamma'(\cdot, t)$ with $\partial D_{1/2}$, $A'(t)$, and $B'(t)$, converges to two distinct points on γ^* as $t \uparrow \omega$.

7.3   Shrinking to a point

Next, we'd like to derive an isoperimetric type estimate for the evolving arc Γ' as described in the last paragraph of the previous section.

For each fixed $t \in [t_1, \omega)$, let β be any embedded curve whose distinct endpoints lie on $\Gamma'(\cdot, t)$ and whose interior is disjoint from $\Gamma'(\cdot, t)$. Let $\mathcal{L}(t)$ be the class of all such curves. Letting L be the length of β, we define

$$G(\beta) = L^2 \left(\frac{1}{A} + 1 \right)$$

and

$$g(t) = \inf \left\{ G(\beta) : \beta \in \mathcal{L}(t) \right\} \, ,$$

where A is the area of the region enclosed by β and $\Gamma'(\cdot, t)$. It is not hard to see that there exists a small $\varepsilon_0 > 0$ such that, whenever $g(t) < \varepsilon_0$, the infimum $g(t)$ is taken on the subclass

$$\mathcal{L}'(t) = \left\{ \beta \in \mathcal{L}'(t) \; : \; \text{the endpoints of } \beta \text{ lie on } \Gamma'(\cdot, t) \cap D_{1/4} \right\} \, .$$

Lemma 7.5 *If $g(t) < \varepsilon_0$, the infimum $g(t)$ is attained in $\mathcal{L}'(t)$. It has constant curvature and is perpendicular to $\Gamma'(\cdot, t)$ at its endpoints.*

Proof: Let $\{\beta_j\}$ be a minimizing sequence of $g(t)$ in $\mathcal{L}'(t)$. Let A_j be the area enclosed by β_j and $\Gamma'(\cdot, t)$. We first consider the minimizing problem:

$$L_j = \inf \left\{ \text{length of } \beta : \beta \text{ is a curve in } \mathcal{L}'(t) \text{ whose enclosed area} \right.$$
$$\left. \text{with } \Gamma'(\cdot, t) \text{ is equal to } A_j \right\}.$$

It is well-known that this constrained minimization problem is attained by a curve β_j' whose endpoints intersect $\Gamma'(\cdot, t)$ at right angles. Arguing as in §5.1, one can show that the interior of β_j' does not touch $\Gamma'(\cdot, t)$. Hence, β_j' is a circular arc or a line segment whose interior is disjoint from $\Gamma'(\cdot, t)$. Replacing the minimizing sequence $\{\beta_j\}$ by $\{\beta_j'\}$, we can argue as before that it subconverges to a minimizer of $g(t)$, which is again a circular arc or a line segment meeting $\Gamma'(\cdot, t)$ at right angles and disjoint from $\Gamma'(\cdot, t)$ at its interior points. $\square$

Now we compute the first and second variations at a minimizer β_0 for a fixed time. Here, the time variable is suppressed. For $\mu \in [-\mu_0, \mu_0], \mu_0 > 0$. Let β_μ be any one-parameter family of admissible curves, $A(\mu)$ and $L(\mu)$ are, respectively, the area enclosed by β_μ and $\Gamma'(\cdot, \mu)$, and the length of β_μ. For simplicity, we assume the curvature of β_0, k_0, is nonzero. The case $k_0 = 0$ can be treated similarly. Using polar coordinates at the center of β_0, which is a circular arc, β_μ is given by the graph of $r = r(\theta, \mu)$ between $\theta_- = \theta_-(\mu)$ and $\theta_+ = \theta_+(\mu)$ and $\beta_0 = \{(r, \theta) : r = |k_0|^{-1}, \theta \in [\theta_-, \theta_+]\}$.

Since the function $r(\theta, 0)$ is constant and β_0 is perpendicular to

$\Gamma'(\cdot, t)$, we have

$$\frac{\partial r}{\partial \theta} = 0 \; , \quad \frac{\partial^2 r}{\partial \theta^2} = 0$$

and

$$\frac{\partial \theta_+}{\partial \mu} = 0 \; , \quad \frac{\partial \theta_-}{\partial \mu} = 0 \; ,$$

at $\mu = 0$. The curvatures of $\Gamma'(\cdot, t)$ at $\theta = \theta_+$ and $\theta = \theta_-$ are given by k'_+ and k'_-, respectively. They can be computed as the graph of

$$\theta = \theta_+(\mu) \text{ and } r = r\big(\theta_+(\mu), \mu\big)$$

or

$$\theta = \theta_-(\mu) \text{ and } r = r\big(\theta_-(\mu), \mu\big) \; ,$$

except with a possible sign change, since we keep the sign convention for k' in the flow (7.10). By a direct computation, we have

$$\left(r \frac{d^2 \theta}{d\mu^2} \right)_+ = -k'_+ \left(\frac{\partial r}{\partial \mu} \right)_+^2$$

and

$$\left(r \frac{d^2 \theta}{d\mu^2} \right)_- = k'_- \left(\frac{\partial r}{\partial \mu} \right)_-^2 \; ,$$

at $\mu = 0$. Here and in the following subscript, " $+$ " or " $-$ " means evaluation at $\theta = \theta_+$ or θ_-.

The computation and result could be easily expressed in terms of the velocity v and the acceleration z of β_μ,

$$v = \frac{\partial r}{\partial \mu} \quad \text{and} \quad z = \frac{\partial^2 r}{\partial \mu^2} \; ,$$

at $\mu = 0$. The length of β_μ is given by

$$L(\mu) = \int_{\theta_-}^{\theta_+} \sqrt{r^2 + \left(\frac{dr}{d\theta} \right)^2} \, d\theta \; .$$

For simplicity, we may assume the region enclosed by β_μ and $\Gamma'(\cdot, t)$ is on the origin side of β_μ. Then,

$$A(\mu) - A(0) = \int_0^\mu \int_{\theta_-(\tau)}^{\theta_+(\tau)} r \frac{\partial r}{\partial \mu} d\theta d\tau .$$

By a straightforward computation, we have,

Lemma 7.6 *At $\mu = 0$, we have*

$$\frac{dL}{d\mu} = \int_{\theta_-}^{\theta_+} v d\theta ,$$

$$\frac{d^2 L}{d\mu^2} = \int_{\theta_-}^{\theta_+} z d\theta + |k_0| \int_{\theta_-}^{\theta_+} \left(\frac{dv}{d\theta}\right)^2 d\theta - \left(k'_+ v_+^2 + k'_- v_-^2\right) ,$$

$$\frac{dA}{d\mu} = \frac{1}{|k_0|} \int_{\theta_-}^{\theta_+} v d\theta , \qquad\qquad and$$

$$\frac{d^2 A}{d\mu^2} = \frac{1}{|k_0|} \int_{\theta_-}^{\theta_+} z d\theta + \int_{\theta_-}^{\theta_+} v^2 d\theta .$$

Since $\log G$ attains its minimum at β_0, we have

$$0 = \frac{d}{d\mu}\Big|_{\mu=0} \log G$$

$$= \left(\frac{2}{L} + \frac{1}{|k_0|(A+1)} - \frac{1}{|k_0|A}\right) \int_{\theta_-}^{\theta_+} v d\theta .$$

One can choose the integral on the right non-zero. Therefore,

$$\frac{2}{L} = \frac{1}{|k_0|A(A+1)} . \tag{7.11}$$

Next, by Lemma 7.6 and (7.11),

$$0 \leqslant \frac{d^2}{d\mu^2}\Big|_{\mu=0} \log G$$

$$= \frac{2}{L}\left[\int_{\theta_-}^{\theta_+} z\,d\theta + |k_0| \int_{\theta_-}^{\theta_+} \left(\frac{dv}{d\theta}\right)^2 d\theta - \left(k'_+ v_+^2 + k'_- v_-^2\right)\right]$$

$$- \frac{2}{L^2}\left(\int_{\theta_-}^{\theta_+} v\,d\theta\right)^2 + \frac{1}{A+1}\left(\frac{1}{|k_0|}\int_{\theta_-}^{\theta_+} z\,d\theta + \int_{\theta_-}^{\theta_+} v^2\,d\theta\right)$$

$$- \frac{1}{(A+1)^2}\left(\frac{1}{|k_0|}\int_{\theta_-}^{\theta_+} v\,d\theta\right)^2 - \frac{1}{A}\left(\frac{1}{|k_0|}\int_{\theta_-}^{\theta_+} z\,d\theta + \int_{\theta_-}^{\theta_+} v^2\,d\theta\right)$$

$$+ \frac{1}{A^2}\left(\frac{1}{|k_0|}\int_{\theta_-}^{\theta_+} v\,d\theta\right)^2$$

$$= \frac{2|k_0|}{L}\int_{\theta_-}^{\theta_+} \left[\left(\frac{dv}{d\theta}\right)^2 - v^2\right]d\theta - \frac{2}{L}\left(k'_+ v_+^2 + k'_- v_-^2\right)$$

$$+ \frac{1}{2|k_0|^2}\frac{4A+1}{A^2(A+1)^2}\left(\int_{\theta_-}^{\theta_+} v\,d\theta\right)^2 .$$

Choosing

$$v = \sqrt{\Phi\big(|k_0|^{-1}\cos\theta, |k_0|^{-1}\sin\theta\big)} ,\ \theta \in [\theta_-, \theta_+] ,$$

we have

$$\left(\frac{dv}{d\theta}\right)^2 \leqslant C_1\left(\frac{1}{|k_0|^2} + 1\right)$$

for some positive constant C_1 depending on Φ. Noting that $|k_0| = (\theta_+ - \theta_-)/L$, we deduce from (7.11) that

$$\frac{2|k_0|}{L}\int_{\theta_-}^{\theta_+} \left(\frac{dv}{d\theta}\right)^2 d\theta \leqslant 2C_1\left[1 + \frac{L^2}{A^2(A+1)^2}\right] .$$

Thus,

$$\frac{2}{L}\left(k'_+ \Phi(x_+, y_+) + k'_- \Phi(x_-, y_-)\right) \leqslant C_2\left(1 + \frac{L^2}{A^2(A+1)^2}\right) \quad (7.12)$$

for some constant C_2 depending on Φ. Here, (x_+, y_+) and (x_-, y_-) are the position vectors of the endpoints of β_0.

Now we can state the isoperimetric type estimate.

Lemma 7.7 *There exists a positive constant δ such that $g(t) \geqslant \delta > 0$ on $[t_1, \omega)$.*

Proof: Fix $\bar{t} \in [t_1, \omega)$ and let β be the minimizer of $g(\bar{t})$. For t close to $\bar{t}$, let β_t be the continuously changing circular arcs or line segments whose endpoints lie on $\Gamma'(\cdot, t)$ and $\beta_{\bar{t}} = \beta$. The time derivative of the length of β_t at $\bar{t}$ is the sum of the negative normal speed of $\Gamma'(\cdot, t)$ at the endpoints of β_t. By noting the orientation of $\Gamma'(\cdot, t)$, the length $L(t)$ of β_t satisfies

$$\frac{dL}{dt}\Big|_{t=\bar{t}} = -\Big(\Phi(x_-, y_-)k'_- + \Psi(x_-, y_-, \theta_-) + \Phi(x_+, y_+)k'_+$$

$$+ \Psi(x_+, y_+, \theta_+ + \pi)\Big) .$$

According to (1.19),

$$\frac{dA}{dt} = -\int_{\gamma'(\cdot, t)} \Big(\Phi(\gamma')k' + \Psi(\gamma', \theta)\Big) ds ,$$

where $\gamma'(\cdot, \bar{t})$ is the portion of $\Gamma'(\cdot, \bar{t})$ which together with β bounds the area $A(\bar{t})$.

Let $(\bar{x}, \bar{y})$ be any fixed point on $\gamma'(\cdot, \bar{t})$. By a direct computation,

$$\frac{d}{dt}\Big|_{t=\bar{t}} \log G = -\frac{2}{L}\Big(\Phi(x_+, y_+)k'_+ + \Phi(x_-, y_-)k'_- - \Psi(x_+, y_+, \theta_+)$$

$$+ \Psi(x_-, y_-, \theta_-)\Big) + \frac{1}{A(A+1)}\int_{\gamma'(\cdot, \bar{t})} (\Phi k' + \Psi) ds$$

$$\geqslant -\frac{2}{L}\Big(\Phi(x_+,y_+)k'_+ + \Phi(x_-,y_-)k'_- - \Psi(x_+,y_+,\theta_+) + \Psi(x_-,y_-,\theta_-)$$

$$+\frac{1}{A(A+1)}\int_{\gamma'(\cdot,\bar{t})}\Phi(\bar{x},\bar{y})k'ds - \frac{1}{A(A+1)}\Big(\int_{\gamma'(\cdot,\bar{t})}\big|\Phi(x,y)$$

$$-\Phi(\bar{x},\bar{y})\big|\big|k'\big|ds + \big|\Psi\big|_{\max}L\big(\Gamma'(\cdot,\bar{t})\big)\Big)\Big),\qquad(7.13$$

where $L\big(\Gamma'(\cdot,\bar{t})\big)$ is the length of $\Gamma'(\cdot,\bar{t})$.

By Lemma 7.3, we know

$$\int_{\gamma'(\cdot,\bar{t})}\big|\Phi(x,y) - \Phi(\bar{x},\bar{y})\big||k'|ds$$

$$\leqslant \quad C_3 \max_{\gamma'(\cdot,\bar{t})}\big|\Phi(x,y) - \Phi(\bar{x},\bar{y})\big| \,,$$

where C_3 only depends on Φ, Ψ, and the initial curve. Without loss of generality, we assume $\theta_- \leqslant 0 \leqslant \theta_+$. Then,

$$\int_{\gamma'(\cdot,\bar{t})}\Phi(\bar{x},\bar{y})k'ds \quad = \quad \int_{\theta_++\pi}^{\theta_-+2\pi}\Phi(\bar{x},\bar{y})d\theta$$

$$= \quad \Phi(\bar{x},\bar{y})\big(\pi - \theta_+ + \theta_-\big)\,.$$

Thus, (7.13) can be written as

$$\frac{d}{dt}\Big|_{t=\bar{t}}\log G \geqslant$$

$$-\frac{2}{L}\Big[\Phi(x_+,y_+)k'_+ + \Phi(x_-,y_-)k'_- - \Psi(x_+,y_+,\theta_+) + \Psi(x_-,y_-,\theta_-)\Big]$$

$$+\frac{1}{A(A+1)}\Phi(\bar{x},\bar{y})\big[\pi - (\theta_+ - \theta_-)\big]$$

$$-\frac{1}{A(A+1)}\Big(C_3 \max_{\gamma'(\cdot,\bar{t})}\big|\Phi(x,y) - \Phi(\bar{x},\bar{y})\big| + \big|\Psi\big|_{\max}L\big(\Gamma'(\cdot,\bar{t})\big)\Big)\,.$$

An application of the mean value theorem shows

$$\frac{d}{dt}\Big|_{t=\bar{t}} \log G \;\geqslant\; -\frac{2}{L}\big(\Phi(x_+, y_+)k'_+ + \Phi(x_-, y_-)k'_-\big) - C_5\Big(1 + \frac{\theta_+ - \theta_-}{L}\Big)$$

$$+\frac{C_4}{A(A+1)}\big[\pi - (\theta_+ - \theta_-)\big] - \frac{C_5}{A(A+1)}L\big(\Gamma'(\cdot, \bar{t})\big) \;.$$

Finally, we use (7.11), (7.12), and $\theta_+ - \theta_- = |k_0|L$ to get

$$\frac{d}{dt}\Big|_{t=\bar{t}} \log G \geqslant \frac{1}{A(A+1)}\Big\{C_4\pi - C_6\big[g(\bar{t}) + L\big(\Gamma'(\cdot, \bar{t})\big) + A\big]\Big\}, \quad (7.14)$$

where C_4 and C_6 depend only on Φ, Ψ, and the initial curve.

Observe that $\Gamma'(\cdot, t)$ converges to a portion of γ^* smoothly away from Q and the total absolute curvature of $\Gamma'(\cdot, t)$ is uniformly bounded. By replacing $\Gamma'(\cdot, t)$ by the portion lying on a sufficiently small neighborhood of the origin, if necessary, we may assume $L\big(\Gamma'(\cdot, t)\big)$ and A are so small that

$$C_6\Big(L\big(\Gamma'(\cdot, \bar{t})\big) + A\Big) < \frac{1}{2}C_4\pi \quad, \qquad \text{for all } \bar{t} \in (t_1, \omega) \;.$$

Hence, by (7.14), for $t < \bar{t}$, and close to $\bar{t}$,

$$g(t) \leqslant G(\beta_t) < G(\beta_{\bar{t}}) = g(\bar{t}) \;.$$

We conclude that there exists $\delta \leqslant \varepsilon_0$ such that $g(t) \geqslant \delta > 0$ for all t in $[t_1, \omega)$. $\qquad\square$

Theorem 7.8 *Let $\gamma(\cdot, t)$ be the solution (7.1) where (7.4) holds. If ω is finite, $\gamma(\cdot, t)$ shrinks to a point as $t \uparrow \omega$.*

Proof: We employ the blow-up argument in §5.2 to (7.10). In case the flow does not shrink to a point, we get a limit flow γ_∞ satisfying

$$\frac{\partial \gamma_\infty}{\partial t} = \Phi(0, 0)k_\infty \boldsymbol{n}' \;.$$

As before, one can show that γ_∞ is uniformly convex and with total absolute curvature π. Now we can follow the proof of Theorem 5.1, using Lemma 7.7 to replace Proposition 5.2, to draw a contradiction. $\square$

7.4   Convergence to a geodesic

In this section, we prove the following theorem, which complements Theorem 7.8.

Theorem 7.9 *Let $\gamma(\cdot,t)$ be the solution of (7.1) where (7.4) holds. If it exists for all $t \geqslant 0$, its curvature converges to zero uniformly as $t \longrightarrow \infty$. Consequently, $\gamma(\cdot,t)$ subconverges to an embedded, closed geodesic of (M,g).*

Lemma 7.10

$$\lim_{t\to\infty} \int_{\gamma(\cdot,t)} k^2(s,t)ds = 0 .$$

Proof: By Lemma 7.1, we have

$$\frac{d}{dt}\int k^2 ds \;=\; \int \left[2k\left(F_{ss} + k^2 F + KF\right) - Fk^3\right]ds \qquad (7.15)$$

$$=\; \int (-2ak_s^2 - 2a_s kk_s + ak^4 + 2aKk^2)ds .$$

Here and in the followings, the integration is over $\gamma(\cdot,t)$.

Since the flow exists for all time, its length $L(t)$ must have a positive lower bound, for otherwise $\gamma(\cdot,t)$ would be contained in a very small geodesic disk and, by (7.10), shrinks to a point in finite time. Consequently, we can find some constant C depending only on a and M such that $\inf k(\cdot,t) \leq C$ for all t. So,

$$\max_{\gamma(\cdot,t)} k^2 \leqslant C^2 + L(t)\int k_s^2 ds .$$

Putting this into (7.15), we have

$$\frac{d}{dt} \int k^2 ds \leqslant - \int ak_s^2 ds + L(t) \int k_s^2 ds \int ak^2 ds + C_1 \int k^2 ds .$$

By Lemma 7.1, for any ε less than $a_{\min}(2L(0)a_{\max})^{-1}$, we can find t_1 such that

$$\int_{\gamma(\cdot,t_1)} k^2 ds < \frac{\varepsilon}{2} \tag{7.16}$$

and

$$\int_{t_1}^{\infty} \int_{\gamma(\cdot,t)} ak^2 < \varepsilon^2 . \tag{7.17}$$

Therefore, as long as the total squared curvature of $\gamma(\cdot, t)$ is less than ε,

$$\frac{d}{dt} \int_{\gamma(\cdot,t)} k^2 ds \leqslant -\frac{a_{\min}}{2} \int_{\gamma(\cdot,t)} k_s^2 ds + C_2 \int_{\gamma(\cdot,t)} k^2 ds \tag{7.18}$$

for some C_2. We claim that

$$\int_{\gamma(\cdot,t)} k^2 < \varepsilon$$

for all $t \geqslant t_1$. For, if on the contrary, there is a first time $t_2 > t_1$ such that

$$\int_{\gamma(\cdot,t_2)} k^2 = \varepsilon ,$$

then, by (7.18), (7.16), and (7.17),

$$\frac{\varepsilon}{2} < \int_{\gamma(\cdot, t_2)} k^2 ds - \int_{\gamma(\cdot, t_1)} k^2 ds$$

$$\leqslant \frac{C_2}{a_{\min}} \int_{t_1}^{t_2} \int ak^2 ds\, dt$$

$$< \frac{C_2 \varepsilon^2}{a_{\min}} \ ,$$

which is impossible after a further restriction on ε. $\square$

Lemma 7.11

$$\lim_{t \longrightarrow \infty} \int_{\gamma(\cdot, t)} k_s^2(\cdot, t) ds = 0 \ .$$

Proof: By Lemma 7.1,

$$\frac{d}{dt} \int k_s^2 ds = \int \left[2k_s \left(F_{ss} + k^2 F + KF \right)_s + kFk_s^2 \right] ds$$

$$= \int \Big[-2ak_{ss}^2 + 4a_{ss}kk_{ss} + 2a_{sss}kk_s + 2a_s k^3 k_s - 2aKkk_{ss}$$

$$+ 7ak^2 k_s^2 \Big] ds \ .$$

By the chain rule,

$$|a_s| \leqslant C_0 \ ,$$

$$|a_{ss}| \leqslant C_0 \left(1 + |k| \right) \ , \qquad\qquad \text{and}$$

$$|a_{sss}| \leqslant C_0 \left(1 + k^2 + |k_s| \right) \ ,$$

where C_0 depends only on a. So,

$$\frac{d}{dt} \int k_s^2 ds$$

$$\leqslant + C_1 \int \left(k^2 + k^4 + |kk_s| + |k^3 k_s| + |kk_s^2| + k^2 k_s^2 \right) ds \qquad (7.19)$$

$$- 2 \int a k_{ss}^2 ds \;,$$

where C_1 only depends on a and M.

Since $L(t)$ is decreasing, there exists a constant C_2 such that

$$\int_{\gamma(\cdot,t)} k_s^2 ds \leqslant C_2 \int_{\gamma(\cdot,t)} k_{ss}^2 ds \;.$$

By the Cauchy-Schwarz inequality,

$$\max_{\gamma(\cdot,t)} k_s^2 \leqslant L(t) \int_{\gamma(\cdot,t)} k_{ss}^2 ds \;.$$

So, for $\varepsilon > 0$,

$$\int k^4 ds \quad \leqslant \quad L(t) \int k^2 ds \int k_s^2 ds \;,$$

$$\int |kk_s| ds \quad \leqslant \quad \varepsilon \int k_s^2 ds + \frac{1}{4\varepsilon} \int k^2 ds \;,$$

$$\int |k^3 k_s| ds \quad \leqslant \quad L(t) \int k^2 ds \int k_{ss}^2 ds + \int k^4 ds \;,$$

$$\int |kk_s^2| ds \quad \leqslant \quad L(t) \int |k| ds \int k_{ss}^2 ds$$

$$\leqslant \quad L(t)^{3/2} \left(\int k^2 ds \right)^{1/2} \left(\int k_{ss}^2 ds \right) \;,$$

and

$$\int k^2 k_s^2 ds \leqslant L(t) \int k^2 ds \int k_{ss}^2 ds \;.$$

Putting all these estimates into (7.19) and using Lemma 7.10, we conclude

$$\frac{d}{dt}\int k_s^2 ds \;\leqslant\; C_3 \int k^2 ds$$

$$\leqslant\; -\frac{C_3}{a_{\min}}\frac{d}{dt}L(t)$$

for all large t. Hence, the limit

$$\lim_{t\longrightarrow\infty}\int k_s^2(\cdot,t)ds$$

exists. In view of (7.18), the limit must, equal to 0.   □

Proof of Theorem 7.9   By Lemmas 7.10 and 7.11, we know that the curvature $k(\cdot,t)$ tends to zero uniformly as $t \to \infty$. On the other hand, by Lemma 7.1, we can express the curvature k as a function on $[0,L(t)] \times [0,\infty)$ and it satisfies a uniformly parabolic equation. By parabolic theory and interpolation, all derivatives of k also tend to zero as $t \to \infty$, as well. So, subconvergence to a closed geodesic follows from the Ascoli-Arzela theorem. Let $\{\gamma(\cdot,t_j)\}$ satisfy

$$\lim_{t_j \longrightarrow \infty}\gamma(\cdot,t_j) = \gamma_\infty(\cdot) \;,$$

where γ_∞ is a closed geodesic of length L_∞. By the strong maximum principle, there exists a positive integer $N \geqslant 1$ such that $\gamma_\infty\big|_{[0,L_\infty/N)}$ is an embedding. Any given narrow tubular neighborhood of γ_∞ is diffeomorphic to $S^1 \times (-1,1)$. $\gamma(\cdot,t_j)$ are contained inside this neighborhood for all large t_j and collapse to γ_∞ as $j \to \infty$. When $N \geqslant 2$, $\gamma(\cdot,t_j)$ would turn around γ_∞ more than once. Since each $\gamma(\cdot,t_j)$ is embedded, this is not possible. So, N must be equal to 1. The proof of Theorem 7.9 is completed.

As an application of Theorems 7.8 and 7.9, we consider the flow on a convex surface M,

$$\frac{\partial \gamma}{\partial t} = \frac{k}{K} \mathbf{N} , \tag{7.20}$$

where K is the Gaussian curvature of M. We consider the Cauchy problem of (7.20), where γ_0 is an embedded, closed curve satisfying

$$\int_{\gamma_0} k_0 ds = 0 .$$

By the Gauss-Bonnet theorem, γ_0 divides M into two regions of total curvature equal to 2π. By Lemma 7.1(v), we know that the flow $\gamma(\cdot, t)$ satisfies

$$\int_{\gamma(\cdot,t)} k(\cdot, t) ds = 0$$

for all t in $[0, \omega)$. Since the flow cannot shrink to a point, for otherwise the total curvature of one of the regions it divides would tend to zero, we conclude from Theorem 7.9 that $\gamma(\cdot, t)$ exists for all time, and so there exists a closed geodesic on (M, g). We have proved:

Corollary 7.12 *There exists an embedded closed geodesic on a closed surface with positive Gaussian curvature.*

One may also look at the standard curve shortening problem on M,

$$\frac{\partial \gamma}{\partial t} = k\mathbf{N} , \quad \gamma(\cdot, 0) = \gamma_0 , \tag{7.21}$$

where γ_0 is an embedded closed curve and M is simply-connected. Then, either $\gamma(\cdot, t)$ shrinks to a point or it subconverges to a geodesic.

Corollary 7.13 *There exist three embedded closed geodesics on any 2–sphere.*

We sketch the proof following Grayson [66]. Let (Σ, Σ_0) be the space of embedded closed curves relative to the point curves Σ_0. It can be shown that there exists a retract of (Σ, Σ_0) onto $(\mathbb{RP}^3 \backslash D^3, \partial)$. So, this space has homology classes h_1, h_2, h_3 of dimensions one, two, and three, respectively. For each $h \in H_*(\Sigma, \Sigma_0)$, one may consider the minmax problem,

$$\lambda(h) = \inf_C \sup_\gamma L(\gamma),$$

where γ belongs to the cycle C representing h. We can solve (7.21) using each $\gamma \in C$ as the initial curve and obtain $\gamma(\cdot, t)$ in the same cycle. By the curve shortening property of the flow and the definition of $\lambda(h)$, the set $\{\gamma : L(\gamma) \geqslant \lambda(h) - \varepsilon\}$, $\varepsilon > 0$, is nesting and is non-empty for all time. Hence, we can find a solution which exists for all time in C. By Theorem 7.11, it subconverges to an embedded closed geodesic of length $\lambda(h)$. By a topological argument, one can show that, if $\lambda(h_i) = \lambda(h_j)$ for some distinct $i, j \in \{1, 2, 3\}$, there are infinitely many embedded closed geodesics. Hence, in any case, there are at least three of them.

Notes

Poincaré's approach to the existence of an embedded closed geodesic on a convex surface was first rigorously justified in Croke [39]. The flow approach (7.20) was proposed in [56], and its long-time existence is established in [92]. Here, our proof is based on [113].

The theorem of three geodesics was first outlined by Lusternik and Schnirelmann. See [66], Ballmann-Thorbergesson-Ziller [20], and Klingenberg [85] for further discussion on the minimax scheme and the topological part of the proof.

Other geometric applications of the CSF or its variants can also be found in Gage [55] and Angenent [16].

Chapter 8

The Non-convex Generalized Curve Shortening Flow

In this chapter, we study the Cauchy problem for the generalized curve shortening flow (GCSF),

$$\frac{\partial \gamma}{\partial t} = |k|^{\sigma-1} k \boldsymbol{n} \ , \ \ \gamma(\cdot, 0) = \gamma_0 \ , \tag{8.1}_\sigma$$

where γ_0 is a C^2-embedded closed curve. We shall investigate whether the flow exists until it shrinks to a point and whether the Grayson convexity theorem still holds. The main difficulty is that the equation is singular\degenerate parabolic. The main result of this chapter is the following almost convexity theorem.

Theorem 8.1 *The maximal solution of* $(8.1)_\sigma$, $\sigma \in (0, 1)$, *converges to a line segment or a point, and its total absolute curvature tends to* 2π *as* $t \uparrow \omega$.

Since the equation becomes singular when $\sigma \in (0, 1)$, the strong maximum principle does not always hold. We need to redo everything carefully. This is done in Sections 1 and 2. Of particular interest is a direct generalization of a result of Angenent-Sapiro-Tannenbaum which gives an upper bound on the number of convex arcs of the

203

flow. Collapsing into a line segment or a point is proved in Section 3. In Sections 4 and 5, we prove the total absolute curvature of the flow tends to 2π.

8.1   Short time existence

We shall always assume the flow is non-convex and, so, the GCSF has to be singular or degenerate. We cannot deduce local existence directly from the results in Chapter 1. Instead, we use an approximation argument.

We first approximate $|k|^{\sigma-1}k$ by

$$F_\delta(k) = \sigma \int_0^k (\delta + s^2)^{\frac{\sigma-1}{2}} ds \ , \quad \text{where } \delta \in (0,1) \ .$$

Consider the Cauchy problem

$$\begin{cases} \dfrac{\partial \gamma}{\partial t} = F_\delta(k)\boldsymbol{n} \ , \\[2ex] \gamma(\cdot,0) = \gamma_0 \ . \end{cases} \tag{8.2}$$

By Proposition 1.2, (8.2) has a maximal solution in $[0, \omega_\delta)$.

By (1.16), the curvature of γ_δ, k_δ, evolves by

$$\frac{\partial k_\delta}{\partial t} = (F_\delta)_{ss} + k_\delta^2 F_\delta \ . \tag{8.3}$$

Let M_0 be the supremum of the curvature of γ_0. It follows from applying the maximum principle to (8.3) that there exist positive constants M and T depending only on M_0 such that

$$|k_\delta(\cdot,t)| \leqslant M \ , \quad \forall t \in [0,T] \ , \quad \forall \delta \in (0,1) \ . \tag{8.4}$$

Lemma 8.2 *Let $\sigma \in (0,1)$. There exists a constant C depending on M_0 and the length of γ_0 such that*

$$\int_{\gamma_\delta(\cdot,t)} \big(F_\delta(k_\delta)\big)_s^2 ds \leqslant \frac{C}{t} \ , \quad t \in [0,T] \ , \quad 0 < \delta < 1 \ .$$

Proof: For simplicity, we set

$$k = k_\delta \ , \ v = F = F_\delta(k_\delta) \ , \ \text{and} \ w = v_s = F'_\delta(k_\delta)(k_\delta)_s \ .$$

We have

$$v_t = F'v_{ss} + F'k^2v$$

and

$$w_t = (F'w_s)_s + \left(\frac{F''}{F'}k^2v + 3kv + F'k^2\right)w \ .$$

We note from (8.4) that $(F')^{-1}F''k^2v + 3kv + F'k^2$ is uniformly bounded on $[0,T]$. Therefore,

$$\frac{1}{2}\frac{d}{dt}\int_{\gamma_\delta(\cdot,t)} w^2 ds \ = \ \int_{\gamma_\delta}(ww_t - w^2Fk)ds$$

$$\leqslant \ -\int_{\gamma_\delta} F'w_s^2 ds + B_1\int_{\gamma_\delta} w^2 ds$$

$$\leqslant \ -\rho_1\int_{\gamma_\delta} w_s^2 ds + B_1\int_{\gamma_\delta} w^2 ds \ ,$$

where ρ_1 and B_1 depend only on M_0. By interpolation,

$$\int_{\gamma_\delta} v_s^2 ds \ \leqslant \ \left(\int_{\gamma_\delta} v^2 ds\right)^{\frac{1}{2}}\left(\int_{\gamma_\delta} v_{ss}^2 ds\right)^{\frac{1}{2}}$$

$$\leqslant \ B_2\big(L(0)\big)^{\frac{1}{2}}\left(\int_{\gamma_\delta} w_s^2 ds\right)^{\frac{1}{2}} \ ,$$

where $L(0)$ is the initial length. Therefore, letting $\rho_2 = \rho_1(L(0)B_2^2)^{-1}$,

$$\frac{1}{2}\frac{d}{dt}\int_{\gamma_\delta} w^2 ds \leqslant -\rho_2\left(\int_{\gamma_\delta} w^2 ds\right)^2 + B_1\int_{\gamma_\delta} w^2 ds \ ,$$

which implies

$$\int_{\gamma_\delta} w^2 ds \leqslant \max \left\{ \frac{1}{\rho_2 t} , \frac{2B_1}{\rho_2} \right\} . \qquad \square$$

Lemma 8.1 implies that the curvature of $\gamma_\delta(\cdot, t)$ is uniformly Hölder continuous on every compact subset of $S^1 \times (0, T]$. Thus, we can find a subsequence $\{\gamma_{\delta_j}(\cdot, t)\}$ which C^2-converges to a solution of (8.1) in every compact subset of $S^1 \times (0, T]$.

Next, by applying the weak comparison principle (§1.2) to (1.5), where F is given by (8.1), we have the **containment principle**: let γ_1 and γ_2 be two GCSFs in $[0, T)$. Suppose that $\gamma_1(\cdot, 0)$ is bounded by $\gamma_2(\cdot, 0)$. Then, $\gamma_1(\cdot, t)$ is bounded by $\gamma_2(\cdot, t)$ for all $t \in (0, T)$. As a consequence of this principle, we obtain the existence and uniqueness of $(8.1)_\sigma$ as well as the finiteness of ω. Summing up, we have the following proposition.

Proposition 8.3 *For every C^2-embedded closed γ_0, the Cauchy problem for the GCSF has a unique, embedded solution in $C(S^1 \times [0, \omega)) \cap \widetilde{C}^{2,\alpha}(S^1 \times (0, \omega))$ for some $\alpha \in (0, 1)$ and ω is finite. The curvature becomes unbounded as $t \uparrow \omega$.*

It remains to show that the flow is embedded. For uniformly parabolic flows, the embeddedness preserving property is a consequence of the strong maximum principle. Although the strong maximum principle is not available now, we can still prove that the flow is embedded by the monotonicity property of the following isoperimetric ratio introduced by Huisken [78]. In fact, we shall see more applications of this ratio in the subsequent sections.

Let $\gamma : [a, b] \times [t_0, t_1] \to \mathbb{R}^2$ be an evolving arc satisfying $(8.1)_\sigma$. Define the extrinsic and intrinsic distance functions d and $\ell : [a, b]^2 \times$

$[t_0, t_1] \to \mathbb{R}$ by

$$d(p, q, t) = |\gamma(p, t) - \gamma(q, t)|$$

and

$$\ell(p, q, t) = \int_p^q ds(\cdot, t) \ .$$

Proposition 8.4 *Suppose d/ℓ attains a local minimum at $(p, q, \bar{t}) \in (a, b)^2 \times (t_0, t_1)$. Then,*

$$\frac{d}{dt}(\frac{d}{\ell})(p, q, \bar{t}) \geq 0 \ , \tag{8.5}$$

and "=" holds if and only if γ is a straight line. Consequently, $\inf(d/\ell)$ is increasing as long as it attains interior minimum.

Proof: Since d/ℓ has a global maximum on the diagonal of $[a, b]^2$, we may assume $p \neq q$ and $s(p) > s(q)$ at $\bar{t}$.

Denote

$$\boldsymbol{\omega} = \frac{\gamma(q, \bar{t}) - \gamma(p, \bar{t})}{|\gamma(q, \bar{t}) - \gamma(p, \bar{t})|}$$

and

$$\boldsymbol{e}_1 = \frac{d}{ds}\gamma(p, \bar{t}) \ , \ \text{and} \ \boldsymbol{e}_2 = \frac{d}{ds}\gamma(q, \bar{t}) \ .$$

By the assumption, we have

$$0 = \frac{d}{ds}\Big|_{s=0}\left(\frac{|\gamma(s(p) + s, \bar{t}) - \gamma(q, \bar{t})|}{\ell + s}\right)$$

$$= \frac{1}{\ell}\langle -\boldsymbol{\omega}, \boldsymbol{e}_1\rangle - \frac{d}{\ell^2} \ , \tag{8.6}$$

$$0 = \frac{d}{ds}\Big|_{s=0}\left(\frac{|\gamma(p, \bar{t}) - \gamma(s(q) + s, \bar{t})|}{\ell - s}\right)$$

$$= \frac{1}{\ell}\langle \boldsymbol{\omega}, \boldsymbol{e}_2\rangle + \frac{d}{\ell^2} \ , \tag{8.7}$$

and

$$0 \leq \frac{d^2}{ds^2}\Big|_{s=0}\left(\frac{|\gamma(s(p)+s,\bar{t})-\gamma(s(q)-s,\bar{t})|}{\ell+2s}\right)$$

$$= \frac{2}{\ell^2}\langle\boldsymbol{\omega},\boldsymbol{e}_1+\boldsymbol{e}_2\rangle + \frac{1}{\ell}\left[\frac{|\boldsymbol{e}_1+\boldsymbol{e}_2|^2}{d} + \langle\boldsymbol{\omega},k(q,\bar{t})\boldsymbol{n}(q,\bar{t})\right.$$

$$\left.-k(p,\bar{t})\boldsymbol{n}(p,\bar{t})\rangle - \frac{\langle\boldsymbol{\omega},\boldsymbol{e}_1+\boldsymbol{e}_2\rangle^2}{d}\right] + \frac{4d}{\ell^3}\ .$$

By (8.6) and (8.7),

$$\frac{2}{\ell^2}\langle\boldsymbol{\omega},\boldsymbol{e}_1+\boldsymbol{e}_2\rangle + \frac{4d}{\ell^3} = 0\ .$$

Noticing that $\boldsymbol{\omega}//\boldsymbol{e}_1+\boldsymbol{e}_2$, we have

$$|\boldsymbol{e}_1+\boldsymbol{e}_2|^2 = \langle\boldsymbol{\omega},\boldsymbol{e}_1+\boldsymbol{e}_2\rangle^2\ .$$

Thus,

$$0 \leq \langle\boldsymbol{\omega},k(q,\bar{t})\boldsymbol{n}(q,\bar{t})-k(p,\bar{t})\boldsymbol{n}(p,\bar{t})\rangle\ . \tag{8.8}$$

By (8.6) and (8.7), again,

$$\langle\boldsymbol{\omega},\boldsymbol{n}(q,\bar{t})\rangle = -\langle\boldsymbol{\omega},\boldsymbol{n}(p,\bar{t})\rangle$$

$$= \sqrt{1-(d/\ell)^2} > 0\ .$$

So,

$$k(q,\bar{t})+k(p,\bar{t}) \geq 0\ . \tag{8.9}$$

Now,

$$\frac{d}{dt}\left(\frac{d}{\ell}\right)(p,q,\bar{t})$$

$$= \frac{1}{\ell}\left(1-\frac{d^2}{\ell^2}\right)^{\frac{1}{2}}\left(|k|^{\sigma-1}k(q,\bar{t})+|k|^{\sigma-1}k(p,\bar{t})\right)+\frac{d}{\ell^2}\int_q^p |k|^{\sigma+1}ds(\cdot,\bar{t})$$

$$\geqslant 0 .$$ $\square$

Remark 8.5 More generally, Proposition 8.4 holds for the flow (1.2) if $F = F(q)$ is odd and parabolic.

In concluding this section, we state two maximum principles for the curvature and the tangent angle of the flow. On where the curvature does not vanish we have, by (1.15) and (1.16),

$$k_t = (|k|^{\sigma-1}k)_{ss} + k^2(|k|^{\sigma-1}k) \tag{8.10}$$

and

$$\theta_t = \sigma|k|^{\sigma-1}\theta_{ss} \tag{8.11}$$

Lemma 8.6 *(a) If $k(p,t_0) \geqslant 0$ for $p \in [a,b]$ and $k(a,t_0) > 0$, $k(b,t_0) > 0$, then, for some small $\varepsilon > 0$, $k(p,t) > 0$ in $[a,b] \times (t_0, t_0 + \varepsilon)$.*

(b) If $\theta(p,t_0) \geqslant \theta_0$, for $p \in [a,b]$ and $\theta(a,t_0) > \theta_0$, $\theta(b,t_0) > \theta_0$, then, for some small $\varepsilon > 0$, $\theta(p,t) > \theta_0$ in $[a,b] \times (t_0, t_0 + \varepsilon)$.

Proof: (a) and (b) can be proved in a similar way. In the following, we present a proof of (a).

First we note that the flow can be approximated by a sequence $\{\gamma_\delta(\cdot,t)\}$, $\delta = \delta_j \to 0$, satisfying (8.2). Its speed $v_\delta = F_\delta(k)$ satisfies

$$\frac{\partial v_\delta}{\partial t} = F_\delta'(k)\frac{\partial^2 v_\delta}{\partial s^2} + F_\delta'(k)k^2 v_\delta. \tag{8.12}$$

Since the curvature of γ_δ is uniformly bounded, we can find constants η and M, $0 < \eta < 1 \leqslant M$, such that

$$F'_\delta(k) \geqslant \eta > 0 \quad \text{and} \quad 0 \leqslant k^2 F'_\delta(k) \leqslant M$$

for all $\delta = \delta_j$.

We shall construct a subsolution of (8.12). Let $s_\delta(p, t)$ be the arc-length along $\gamma_\delta(\cdot, t)$ from a to p. By the commutation relation $[\partial/\partial t, \partial/\partial s] = k v_\delta \partial/\partial s$, we have

$$\frac{\partial}{\partial s}\left(\frac{\partial s_\delta}{\partial t}\right) = -k v_\delta ,$$

which is uniformly bounded. Set

$$w_\delta(p, t) = \beta \, \Gamma(s_\delta(p, t) + \xi(t - t_0) + 1, \eta(t - t_0)) ,$$

where ξ and β are constants to be specified later and

$$\Gamma(x, \tau) = \frac{1}{\sqrt{\tau}} \exp\left(-\frac{x^2}{4\tau}\right)$$

is the heat kernel. Notice that $\Gamma_x < 0$ for $x > 0$ and $\Gamma_{xx} > 0$ for $(2\tau)^{1/2}$.

Choose $\xi > -\inf\{\partial s_\delta/\partial t : (p, t) \in [a, b] \times [t_0, t_0 + \varepsilon]\}$ where $\varepsilon > 0$ is so small that $(1 + \xi\varepsilon)^2 > 2\varepsilon$. It follows that

$$
\begin{aligned}
(s_\delta(p, t) + \xi(t - t_0) + 1)^2 &\geqslant (1 + \xi\varepsilon)^2 \\
&\geqslant 2\eta(t - t_0)
\end{aligned}
$$

and

$$
\begin{aligned}
\frac{\partial w_\delta}{\partial t} &= \eta \, \beta \, \Gamma_\tau + \beta\left(\xi + \frac{\partial s_\delta}{\partial t}\right)\Gamma_x \\
&= \eta \frac{\partial^2 w_\delta}{\partial s^2} + \beta\left(\xi + \frac{\partial s_\delta}{\partial t}\right)\Gamma_x \\
&\leqslant F'_\delta(k)\frac{\partial^2 w_\delta}{\partial s^2} + F'_\delta(k)k^2 w_\delta.
\end{aligned}
$$

We conclude that w_δ is a subsolution of (8.12) in $[a, b] \times [t_0, t_0 + \varepsilon)$ for all δ_j, and w_δ vanishes identically at $t = t_0$. If we further choose ε and β small (if necessary), we may assume that $w_\delta \leqslant v_\delta$ on the lateral boundary of $[a, b] \times [t_0, t_0 + \varepsilon]$. By the maximum principle, $w_\delta \leqslant v_\delta$ in $[a, b] \times [t_0, t_0 + \varepsilon]$. Letting $j \to \infty$, we conclude $v \geqslant w_0 > 0$ in $[a, b] \times [t_0, t_0 + \varepsilon]$. □

8.2   The number of convex arcs

Let $\gamma(\cdot, t)$ be a maximal solution of $(8.1)_\sigma$ in $(0, \omega)$. An arc β of $\gamma(\cdot, t)$ is called a **convex arc** if it is the image of a maximal arc $[a, b] \subseteq S^1$ on which $k(\cdot, t) \geqslant 0$ and $k(p, t) > 0$ for some $p \in (a, b)$. An arc β is called a **concave arc** if it is the image of $(a, b) \subseteq S^1$ on which $k(\cdot, t) \leqslant 0$ and $k(p, t) < 0$ for $p \in (a, a + \varepsilon) \bigcup (b - \varepsilon, b)$ for some $\varepsilon > 0$.

First, we show that the number of convex\concave arcs does not increase in time.

Lemma 8.7 *For any (p_0, t_0) with $k(p_0, t_0) \neq 0$, there exists a continuous map: $p : [0, t_0] \longrightarrow S^1$ such that $p(t_0) = p_0$ and $k(p(t), t) \neq 0$ for all t.*

Proof: Without loss of generality, we assume $k(p_0, t_0) > 0$. Consider the set $\{(p, t) \in S^1 \times [0, t_0] : k(p, t) > 0\}$ and let U be its connected component containing (p_0, t_0). It is sufficient to show that $U \bigcap (S^1 \times \{0\})$ is non-empty.

Suppose this is not true. Then $e^{-ct}k$ has a positive maximum at (p_1, t_1) in U where $t_1 > 0$. If we choose $c > \max_U k^{\sigma+1}$, it follows from (8.10) that $0 \leqslant (e^{-ct}k)_t \leqslant (k^{\sigma+1} - c)e^{-ct}k < 0$ at (p_1, t_1). But this is impossible. □

¿From this lemma, we immediately conclude that, if γ_0 is composed of finitely convex and concave arcs, then the number of convex\concave arcs does not increase in time.

Next, we present a rather suprising feature for $(8.1)_\sigma$, $(0 < \sigma < 1)$, due to Angenet-Sapiro-Tannenbaum [19].

Proposition 8.8 *Let K be the total absolute curvature of γ_0. Then, the number of convex arcs of any solution of $(8.1)_\sigma$, $0 < \sigma < 1$, at time t_* does not exceed*

$$K \max \left\{ \frac{1 + CL^2(0)}{t_*^{\frac{1}{1-\sigma}}} \ , \ \frac{2}{\pi} \right\} , \qquad (8.13)$$

where C is an absolute constant.

When $\sigma = 1$, we know that the flow becomes analytic at an instant, and so it has finitely many convex and concave arcs. However, no a priori estimate like (8.13) is known.

By approximating the initial curve by analytic curves with finite inflection points, it suffices to derive (8.13) assuming there are finitely many convex\concave arcs in $[0, \omega)$. Let $\beta = \{\gamma(p, t_*) : p_- \leqslant p \leqslant p_+\}$ be a convex arc. There are intervals $(p_- - \varepsilon, p_-)$ and $(p_+, p_+ + \varepsilon)$ on which $k(\cdot, t_*) < 0$. Set

$$U_- = \text{the connected component of } \{k \neq 0\}$$
$$\text{containing the segment } (p_- - \varepsilon, p_-) \times \{t_*\} ,$$

and

$$U_+ = \text{the connected component of } \{k \neq 0\} \text{ containing}$$
$$\text{the segment } (p_+, p_+ + \varepsilon) \times \{t_*\}.$$

By Lemma 8.7, for each $t \in [0, t_*]$, there exist $(p, t) \in U_-$ and $(q, t) \in U_+$. So, we can define

$$p_-(t) = \sup\{p : (p, t) \in U_-\} \qquad \text{and}$$
$$p_+(t) = \inf\{p : (p, t) \in U_+\} .$$

Let $\beta(\cdot, t)$ be the image of $[p_-(t), p_+(t)]$ under $\gamma(\cdot, t)$. We call $\beta(\cdot, t), t \in [0, t_*]$ the **history** of the arc $\beta = \beta(\cdot, t_*)$.

Set

$$q_-(t) = \inf\{p > p_-(t) : k(p, t) > 0\} \qquad \text{and}$$

$$q_+(t) = \sup\{p < p_+(t) : k(p, t) > 0\}.$$

Lemma 8.9 $k(p, t) = 0$ *for all* $p \in [p_-(t), q_-(t)] \bigcup [q_+(t), p_+(t)]$.

Proof: Suppose $k(p_0, t_0) < 0$ for some (p_0, t_0), $p_0 \in (p_-(t_0), q_-(t_0))$. By Lemma 8.6, $k < 0$ in $[p_-(t_0) - \varepsilon, p_0] \times (t_0, t_0 + \varepsilon)$ for small $\varepsilon > 0$. We can connect (p_0, t_0) to $(p_-(t_0) - \varepsilon, t_0)$ in $\{k \neq 0\}$. In other words, (p_0, t_0) belongs to U_-. The contradiction holds. A similar argument shows that k vanishes on $[q_+(t), p_+(t)]$. $\square$

Lemma 8.10 *We have*

$$\liminf_{t \to t_0} p_-(t) \geqslant p_-(t_0) \ \text{and} \ \limsup_{t \to t_0} p_-(t) \leqslant q_-(t_0),$$

and

$$\limsup_{t \to t_0} p_+(t) \leqslant p_+(t_0) \ \text{and} \ \liminf_{t \to t_0} p_+(t) \geqslant q_+(t_0).$$

Proof: We only prove the first two cases. For small $\varepsilon > 0$, $(p_-(t_0) - \varepsilon, t_0) \subseteq U_-$. Thus,

$$k(p_-(t_0) - \varepsilon, t) < 0, \qquad \forall t \in (t_0 - \delta, t_0 + \delta),$$

where δ is a small number. Since $p_-(t) \geqslant p_-(t_0) - \varepsilon$ for all $t \in (t_0 - \delta, t_0 + \delta)$, by the definition of $p_-(t)$, the first inequality holds.

Next, by the definition of q_-, $k(q_-(t_0) + \varepsilon, t_0) > 0$ for some small ε. By continuity, $k(q_-(t_0) + \varepsilon, t) > 0$ for t in $|t - t_0| < \delta$, δ small. We claim that $p_-(t) \leqslant q_-(t_0) + \varepsilon$ for all $t \in (t_0 - \delta, t_0 + \delta)$.

Let Γ be the segment $\{q_-(t_0) + \varepsilon\} \times (t_0 - \delta, t_0 + \delta)$. Suppose that $p_-(t_1) > q_-(t_0) + \varepsilon$ for some $t_1 \in (t_0 - \delta, t_0 + \delta)$. There is some $p_1 \in (q_-(t_0) + \varepsilon, p_-(t_1))$ such that $(p_1, t_1) \in U_-$. Applying Lemma 8.7 to $[t_1, t_*]$, we can connect (p_1, t_1) to $(p_-(t_*) - \varepsilon, t_*)$ by a path Γ_1 in $U_- \cap (S^1 \times [t_1, t_*])$. There is also a continuous function $u = p^{(2)}(t), 0 \leqslant t \leqslant t_1, u_1 = p^{(2)}(t_1)$, whose graph Γ_2 lies in $U_- \cap (S^1 \times [0, t])$. The paths Γ_1 and Γ_2 together form a path in U_- connecting $(p_-(t_*) - \varepsilon, t_*)$ to the bottom $S^1 \times \{0\}$. Since $k > 0$ on Γ, this path must be disjoint from Γ. Consequently, $p > q_-(t_0) + \varepsilon$ for any point (p, t) in this path for $t \in (t_0 - \delta, t_0 + \delta)$. But this is in conflict with the definition of $p_-(t_0)$ and $q_-(t_0)$. $\square$

Set

$$\Sigma_t = \{\theta(p, t) : p_-(t) \leqslant p \leqslant p_+(t)\}.$$

Lemma 8.11 Σ_t *is strictly nesting in time.*

Proof: Let $\theta_-(t) = \theta(p_-(t), t)$ and $\theta_+(t) = \theta(p_+(t), t)$. We are going to show that $\theta_-(t)$ is strictly increasing and $\theta_+(t)$ strictly decreasing.

Since $k(\cdot, t) = 0$ on $[p_-(t), q_-(t)], \theta_-(t) = \theta(p, t)$ for all $p \in [p_-(t), q_-(t)]$. Together with Lemma 8.10, we know that $\theta_-(t)$ is continuous. At any fixed $t_0 \in (0, t_*)$, there is $\varepsilon > 0$ such that $k(\cdot, t_0) < 0$ on $[p_-(t_0) - \varepsilon, p_-(t_0))$ and $k(\cdot, t_0) > 0$ on $(q_-(t_0), q_-(t_0) + \varepsilon]$. Thus, $\theta(p, t_0) \geqslant \theta_-(t_0)$ on $[p_-(t_0) - \varepsilon, q_-(t_0) + \varepsilon]$ and the inequality is strict at the endpoints of this interval. By Lemma 8.6, we have $\theta(p, t) > \theta_-(t_0)$ on $[p_-(t_0) - \varepsilon, q_-(t_0) + \varepsilon] \times (t_0, t_0 + \delta)$ for some small $\delta > 0$. By Lemma 8.10, $p_-(t)$ lies between $p_-(t_0) - \varepsilon$ and $q_-(t_0) + \varepsilon$ for t close to t_0. So, $\theta_-(t) > \theta_-(t_0)$ for $t \in (t_0, t_0 + \delta)$. We have shown that $\theta_-(t)$ is strictly increasing in t. Similarly, one can show that $\theta_+(t)$ is strictly decreasing. $\square$

Now we can prove Proposition 8.8.

Let

$$\Delta\theta(t) = \sup\{|\theta(p_1, t) - \theta(p_2, t)| : p_-(t) \leqslant p_1, p_2 \leqslant p_+(t)\} .$$

We first show that

$$\Delta\theta(0) \geqslant \min\left\{\frac{t_*^{\frac{1}{1-\sigma}}}{1 + CL(0)^2}, \frac{\pi}{2}\right\} \qquad (8.14)$$

for some constant C.

Assume that $\Delta\theta(0) < \pi/2$. We let $\Sigma_0 \subseteq [-\Delta\theta(0)/2, \Delta\theta(0)/2]$. On any strictly convex\concave part of β_t, we can use the tangent angle to parametrize the flow. The resulting equation for $v = |k|^{\sigma-1}k$ is given by

$$\frac{\partial v}{\partial t} = \sigma|v|^{\frac{1+\sigma}{\sigma}}(v_{\theta\theta} + v) . \qquad (8.15)$$

Let

$$\bar{v} = A(t)\cos\left(\frac{3\pi\theta}{4\Delta\theta(0)}\right) ,$$

where

$$A(t) = \left[A(0)^{-\frac{1+\sigma}{\sigma}} + \frac{Ct}{(\Delta\theta(0))^2}\right]^{-\frac{\sigma}{1+\sigma}} .$$

We have

$$\frac{\partial \bar{v}}{\partial t} = -\frac{\sigma}{1+\sigma}A(t)^{1+\frac{\sigma+1}{\sigma}}\frac{C}{(\Delta\theta(0))^2}\cos\left(\frac{3\pi\theta}{4\Delta\theta(0)}\right)$$

$$\geqslant -\sigma A(t)^{1+\frac{\sigma+1}{\sigma}}\left(\cos\left(\frac{3\pi\theta}{4\Delta\theta(0)}\right)\right)^{1+\frac{\sigma+1}{\sigma}}\left[\left(\frac{3\pi}{4}\right)^2 - (\Delta\theta(0))^2\right]\frac{1}{(\Delta\theta(0))^2}$$

$$= \sigma|\bar{v}|^{\frac{1+\sigma}{\sigma}}(\bar{v}_{\theta\theta} + v)$$

,

provided C is sufficiently small. So $\bar{v}$ is a supersolution of (8.15). By the maximum principle, we obtain,

$$|v(\theta, t)| \leqslant \frac{C(\Delta\theta(0))^{\frac{2\sigma}{1+\sigma}}}{t^{\frac{\sigma}{1+\sigma}}} \qquad (8.16)$$

(choose $A(0) = \infty$) , and

$$|v(\theta, t)| \leqslant \left[\overline{C}^{-\frac{1+\sigma}{\sigma}} + \frac{Ct}{(\triangle\theta(0))^2} \sup_{\beta_0} |v|^{\frac{1+\sigma}{\sigma}} \right]^{-\frac{\sigma}{1+\sigma}} \sup_{\beta_0} |v| \qquad (8.17)$$

(choose $A(0) = \overline{C} \sup\limits_{\beta_0} |v|$ and $\overline{C} = (\cos\frac{3\pi}{8})^{-1}$.) To explain how the maximum principle is applicable, we observe that each β_t is the union of a finite number of convex\concave arcs. On each arc, the tangent angle θ is single-valued and takes values in $(-\triangle\theta(0)/2, \triangle\theta(0)/2)$ by Lemma 8.11. This image interval changes in time, but v always vanishes at its endpoints. On the other hand, $\overline{v}$ has a uniformly positive lower bound in $[-\triangle\theta(0)/2, \triangle\theta(0)/2]$ for $t \in [0, t_*]$. Hence, we can apply the maximum principle.

Set

$$\begin{aligned} t &\equiv T_{\frac{1}{2}}(\sup_{\beta_0} |v|, \ \triangle\theta(0)) \\ &= 2^{\frac{1+\sigma}{\sigma}} \frac{(\triangle\theta(0))^2}{C(\sup\limits_{\beta_0} |v|)^{\frac{1+\sigma}{\sigma}}} \ . \end{aligned}$$

We have

$$\sup_{\beta_{T_{1/2}}} |v| \leqslant \frac{1}{2} \sup_{\beta_0} |v| \ .$$

By the same argument, we get

$$\sup_{\beta_{t+T_{1/2}(t)}} |v| \leqslant \frac{1}{2} \sup_{\beta_t} |v| \ ,$$

for $T_{\frac{1}{2}}(t) = T_{\frac{1}{2}}(\sup\limits_{\beta_t} |v|, \ \triangle\theta(t))$.

We now define an increasing sequence $\{t_j\}_{j=0}^{\infty}$ inductively by setting

$$t_{j+1} = t_j + T_{\frac{1}{2}}(\sup_{\beta_{t_j}} |v| \ , \ \triangle\theta(t_j)) \ .$$

We have

$$\sup_{\beta_{t_{j+1}}} |v| \leqslant \frac{1}{2} \sup_{\beta_{t_j}} |v| \; .$$

Since

$$\Delta\theta(t_j) \leqslant \int_{\beta_{t_j}} |k| ds(\cdot, t_j)$$

$$\leqslant (\sup_{\beta_{t_j}} |v|)^{\frac{1}{\sigma}} \cdot L(t_j) \; ,$$

we have

$$t_{j+1} - t_j \leqslant \left(\frac{2^{\frac{1+\sigma}{\sigma}} L^2}{C}\right) \left(\sup_{\beta_{t_j}} |v|\right)^{\frac{2}{\sigma} - \frac{1+\sigma}{\sigma}}$$

$$\leqslant \left(\frac{2^{\frac{1+\sigma}{\sigma}} L^2}{C}\right) \left(2^{-j} \sup_{\beta_0} |v|\right)^{\frac{1-\sigma}{\sigma}} \; .$$

By (8.16),

$$t_\infty = \lim_{j \to \infty} t_j \leqslant t_0 + \left(\frac{2^{\frac{1+\sigma}{\sigma}} L^2}{C}\right) \left(\sum_0^\infty 2^{\frac{\sigma-1}{\sigma} j}\right) \left(\sup_{\beta_{t_0}} |v|\right)^{\frac{1-\sigma}{\sigma}}$$

$$\leqslant t_0 + \frac{CL^2 (\Delta\theta(0))^{\frac{2(1-\sigma)}{1+\sigma}}}{t_0^{\frac{1-\sigma}{1+\sigma}}} \; .$$

Choosing $t_0 = (\Delta\theta(0))^{1-\sigma}$, we deduce

$$t_\infty \leqslant (1 + CL^2)(\Delta\theta(0))^{1-\sigma} \; .$$

As the arc $\beta = \beta_{t_*}$ is not flat, we must have $t_\infty \geqslant t_*$. Hence,

$$\Delta\theta(0) \geqslant \frac{t_*^{\frac{1}{1-\sigma}}}{1 + CL^2} \; .$$

To finish the proof, we note that disjoint convex arcs have disjoint histories. So, the number of convex arcs at t_* does not exceed

$$K \max \left\{ \frac{1 + CL^2}{t_*^{\frac{1}{1-\sigma}}}, \frac{2}{\pi} \right\} ,$$

where K is the total absolute curvature of the initial curve.   □

In view of Proposition 8.8, without loss of generality, we may assume that the numbers of convex and concave arcs of the flow are always constant. We call an evolving arc β an **evolving convex (or concave) arc** if, for each $t \in [0, \omega)$, $\beta(\cdot, t)$ is a convex\concave arc and the family $\{\beta(\cdot, t') : 0 \leqslant t' < t\}$ is the history of $\beta(\cdot, t)$. Now the flow can be decomposed into a finite union of evolving convex\concave arcs. Notice by Lemma 8.11 the spherical images of evolving convex\concave arcs are strictly nesting in time.

8.3   The limit curve

First of all, by comparing the flow with an evolving circle, one can argue as in the proof of Proposition 1.10 that

$$\gamma(\cdot, t) \subseteq N_{Ct^{\frac{1}{1+\sigma}}}(\gamma_0).$$

As before, the curve $\gamma(\cdot, t)$ converges to a limit curve γ^* in the Hausdorff metric. In this section, we shall show that γ^* is either a single point or a line segment.

In the previous chapters, we have seen that counting the intersection points of two evolving curves is very useful in understanding the behaviour of the flow near the singularity. We now consider this issue for (8.1).

Given two embedded but not necessarily closed C^2-curves γ_1 and γ_2, we count their intersections by the following method: (i) if γ_1 and γ_2 intersect at a point transverally, then count the number of intersections in a small neighborhood of this point, to be 1; and (ii) if γ_1 and γ_2 do not intersect transversally at a point, we choose a coordinate system on a suitable small tubular neighborhood along one of the curves such that this neighborhood contains the point and the restrictions of the curves in this neighborhood are expressed as

graphs of two functions u_1 and u_2. We count the number of intersections in the neighborhood to be the number of sign changes of
$u_1 - u_2$. The **number of crossing** of γ_1 and γ_2 is the sum of the
number of these intersections.

By Lemma 8.7, which only uses the weak maximum principle,
we have:

Proposition 8.12 *Let γ_1 and $\gamma_2 : [0,1] \times [0,T) \to \mathbb{R}$ be two evolving
arcs of (8.1) which satisfy*

$$\partial \gamma_1(\cdot, t) \bigcap \gamma_2(\cdot, t) = \partial \gamma_2(\cdot, t) \bigcap \gamma_1(\cdot, t) = \phi$$

*for all $t \in (0, T)$. Then, the number of crossing of $\gamma_1(\cdot, t)$ and $\gamma_2(\cdot, t)$
is non-increasing in t.*

We point out that this proposition does not assert the number
of crossing is finite, or is decreased by tangency of intersection.

Theorem 8.13 *Let $\gamma(\cdot, t)$ be a maximal solution of (8.1). There
exists finitely many points $\{Q_1, \cdots, Q_m\}$ on the limit curve γ^* such
that $\gamma^* \setminus \{Q_1, \cdots, Q_m\}$ consists of C^2-arcs. Away from $Q_j s$, $\gamma(\cdot, t)$
converges to γ^* in C^2-norm.*

Proof: We closely follow the notation and the proof of Theorem
6.4. As before, $\{\boldsymbol{K}_t\}$, the push-forward of $|k(p, t)|dp$, are uniformly
bounded Borel measures in $\mathbb{R}^2$, by Lemma 8.11. We can select a
sequence $t_j \uparrow \omega$ such that $\{\boldsymbol{K}_{t_j}\}$ converges weakly to a limit measure
$\boldsymbol{K}$,

$$\boldsymbol{K} = \boldsymbol{K}_c + \sum_i K_i \delta_{Q_i},$$

where $\boldsymbol{K}_c = 0$ at points, $\{Q_i\}$ is countable, and $K_1 \geqslant K_2 \geqslant \cdots > 0$
with $\sum K_i$ bounded by the total absolute curvature of the initial
curve. Let m be the first integer for which $K_{m+1} < \pi$. Then, for any

$P \in \gamma^* \setminus \{Q_1, \cdots, Q_m\}$, there is an $\varepsilon > 0$ such that $K(D_\varepsilon(P)) < \pi$. After throwing away the first few t_js, we may assume there exists $\beta > 0$ such that

$$\boldsymbol{K}_{t_j}(D_\varepsilon(P)) \leqslant \pi - \beta \qquad (8.18)$$

for all j.

By passing to a subsequence, if necessary, we can find a fixed number N such that there are exactly N components in $\gamma(\cdot, t_j) \cap D_\varepsilon(P)$ which also intersect $D_{\varepsilon/2}(P)$, and these components are graphs of uniformly Lipschitz continuous functions $y_{j,1}(x), \cdots, y_{j,N}(x)$ over some fixed interval on the x-axis. For simplicity, let's take P to be the origin. We can find $\xi, \eta > 0$ small such that the two line segments $\ell_\pm = [-\xi, \xi] \times \{\pm\eta\}$ lie inside $D_{\varepsilon/2}(P)$ and are disjoint from $\gamma(\cdot, t)$ for all $t \in [t_{j_0}, \omega)$ where t_{j_0} is close to ω. Any line segment Γ connecting ℓ_+ and ℓ_- is a stationary solution of (8.1). By Proposition 8.12, the number of crossing of Γ and $\gamma(\cdot, t)$ is non-increasing in time. When the slope of Γ is steeper than those of $y_{j_0,1}, \cdots, y_{j_0,N}, \gamma(\cdot, t)$ will have exactly N crossings with Γ for all $t \in [t_{j_0}, \omega)$. So, $\gamma(\cdot, t) \cap [-\xi/2, \xi/2] \times [-\eta, \eta]$ consists of exactly N many curves which are graphs of uniformly Lipschitz continuous functions $y_1(x, t), \cdots, y_N(x, t)$ whose Lipschitz constants only depend on β.

Next, we derive a uniform C^2-bound for these functions in $[-\xi/2 + \delta, \xi/2 - \delta] \times [t_{j_0} + \delta, \omega)$ for small $\delta > 0$. Denote by φ a fixed travelling wave solution of unit speed of (8.1) which is non-negative and convex over some interval $(-a, a)$, $a \leqslant \infty$, and let

$$v(x, t) = c \, \varphi\left(\frac{x - \bar{x}}{c}\right) + \left(\frac{1}{c}\right)^\sigma t + \bar{y} ,$$

where $c > 0$ and $(\bar{x}, \bar{y}) \in \mathbb{R}^2$ (see §2.1). Let $y = y_i$ for some i and $(x_0, t_0) \in [-\xi/2 + \delta, \xi/2 - \delta] \times [t_{j_0} + \delta, \omega)$. We can choose $\bar{x}, \bar{y}$

depending on c such that

$$y(x_0, t_0) = v(x_0, t_0)$$

and

$$y_x(x_0, t_0) = v_x(x_0, t_0) .$$

Accordingly, we choose c small (depending on δ and η) such that $|v_x(x, t_{j_0})|$ is greater than the uniform gradient bound of y in $[-\xi/2, \xi/2] \times [t_{j_0}, \omega)$ whenever $|v(x, t_{j_0})| \leqslant \eta$. We also require that $v(x, t_{j_0}) > \eta$ for $|x| \geqslant \xi/2$. Under these conditions, the function $y(x, t_{j_0}) - v(x, t_{j_0})$ has exactly two simple zeroes in $[-\xi/2, \xi/2]$ and

$$y(\pm\frac{\xi}{2}, t) - v(\pm\frac{\xi}{2}, t) < 0, \qquad \forall\, t \in [t_{j_0}, \omega) .$$

On the other hand, $w = y - v$ satisfies a uniformly parabolic equation of the form

$$w_t = a(x, t)w_{xx} + b(x, t)w_x ,$$

because v is uniformly convex. By the Strum oscillation theorem, $w(\cdot, t_0)$ has no other zero other than the one at x_0. So,

$$y_{xx}(x_0, t_0) \leqslant \frac{1}{c}\, \varphi_{xx}\left(\frac{x_0 - \overline{x}}{c}\right) .$$

A similar argument establishes a lower bound for y_{xx}. We have proved that $y_i (1 \leqslant i \leqslant N)$ are C^2-uniformly bounded in $[-\xi/2 + \delta, \xi/2 - \delta] \times [t_{j_0} + \delta, \omega)$.

Now we may apply a regularity result of DiBennedetto on porous medium equations (§4.15 in DiBenedetto [43]) to the evolution equation of $(y_i)_{xx}$ (see (1.3)) to conclude that each $(y_i)_{xx}$ is equicontinuous in $[-\xi/2 + 2\delta, \xi/2 - 2\delta] \times [t_{j_0} + 2\delta, \omega)$. Thus, y_i converges to γ^* in C^2-norm around P. $\square$

Next, we shall show that γ^* in fact consists of line segments. Suppose there is a non-inflection point on the regular part of γ^*. We shall see how this will lead to a contradiction.

Lemma 8.14 *Let P be a non-inflection point. There exist $\varepsilon_0 > 0$ and $t_o \in [0, \omega)$ such that $D_\varepsilon(P) \bigcap \gamma(\cdot, t)$ is a connected arc for all t in $[t_0, \omega)$.*

Proof: Near the point P $\gamma(\cdot, t)$ and γ^* are C^2 with non-zero curvature. Hence, the lemma follows from the strong maximum principle. $\square$

Proposition 8.15 γ^* *consists of, at most, a finite number of line segments.*

Proof: Suppose on the contrary there are two regular non-inflection points $\overline{P}$ and $\overline{Q}$ on $\gamma^* \setminus \{Q_1, \cdots, Q_m\}$. By Theorem 8.13, there exist $a, b \in S^1$ such that $\gamma(a, t)$ and $\gamma(b, t)$ tend to $\overline{P}$ and $\overline{Q}$, respectively, as $t \uparrow \omega$. By Lemma 8.14, we can find $D_\varepsilon(\overline{P})$ and $D_\varepsilon(\overline{Q})$ such that $\gamma(\cdot, t) \bigcap D_\varepsilon(\overline{P})$ and $\gamma(\cdot, t) \bigcap D_\varepsilon(\overline{Q})$ contain a single connected component of $\gamma(\cdot, t)$ for all t close to ω. In particular, we have

$$\text{dist}(\gamma(a, t), \gamma(\cdot, t) \setminus D_\varepsilon(\overline{P})) \geqslant \delta$$

and

$$\text{dist}(\gamma(b, t), \gamma(\cdot, t) \setminus D_\varepsilon(\overline{Q})) \geqslant \delta, \tag{8.19}$$

for some δ. Let $p(t)$ be given by

$$|k(p(t), t)| = \max_{\gamma(\cdot, t)} |k(\cdot, t)|.$$

Since $|k(p(t), t)| \to \infty$ as $t \uparrow \omega$, there exists $\{t_j\}$ such that

$$|k(p, t)| \leqslant |k(p(t_j), t_j)|, \quad \forall (p, t) \in S^1 \times [0, t_j].$$

Without loss of generality, we may assume

$$a < p(t_j) < b, \quad \forall j. \tag{8.20}$$

We consider the rescaling of the blow-up sequence $\{\gamma(\cdot, t_j)\}$ as we did in Chapter 5 by setting

$$\gamma_j(p, t) = \frac{\gamma(p(t_j) + \varepsilon_j p, t_j + \varepsilon_j^{1+\sigma} t) - \gamma(p(t_j), t_j)}{\varepsilon_j},$$

where $(p, t) \in \mathbb{R} \times [-t_j \, \varepsilon_j^{-1-\sigma}, (\omega - t_j) \, \varepsilon_j^{-1-\sigma})$ and

$$\varepsilon_j = |k(p(t_j), t_j)|^{-1}.$$

Each γ_j satisfies (8.1) with $|k_j(p, t)| \leqslant 1 = |k_j(0, 0)|$. By the regularity result of DiBenedetto [43], we can choose a subsequence, still denoted by $\{\gamma_j\}$, which converges to some γ_∞ in C^2-norm in every compact subset of $\mathbb{R} \times (-\infty, 0]$. The limit γ_∞ solves (8.1) with curvature k_∞ satisfying $|k_\infty| \leqslant |k_\infty(0, 0)| = 1$. It is also clear that γ_∞ is complete and noncompact. We claim that it is also convex and embedded.

The number of convex\concave arcs of γ_∞ is finite, and is non-increasing in time by Lemma 8.7. Therefore, we may assume that the number of convex arcs is constant for $t \in (-\infty, -M], M > 0$. It suffices to show $\gamma_\infty(\cdot, t)$ is convex in $(-\infty, -M]$.

Supposing not, then we can find p, q lying on two adjacent arcs of $\gamma_\infty(\cdot, -M)$ such that $k_\infty(p, -M) < 0, k_\infty(q, -M) > 0$, and $p < q$. Let's denote by β_- the evolving concave arc containing $\gamma(p, -M)$ and by β_+ the evolving convex arc containing $\gamma(q, -M)$. For each $t \in (-\infty, -M]$ we let $\theta_-(t)$ be the tangent angle of the inflection point (or spot) which separates β_- and β_+. By Lemma 8.11, $\theta_-(t)$ is strictly increasing in t. So,

$$\theta_-(-M) - \theta_-(-M - 1) \geqslant 2\varepsilon_0 > 0$$

for some ε_0. In other words, the minimum tangent angle of the two adjacent arcs increases at least $2\varepsilon_0$ as time t goes from $-M-1$ to M.

For large j, $\gamma_j(p, -M)$ belongs to an evolving concave arc $\beta_-^{(j)}$ of $\gamma_j(\cdot, t)$ and $\gamma_j(q, -M)$ belongs to an evolving convex arc $\beta_+^{(j)}$ of $\gamma_j(\cdot, t)$. Though $\beta_-^{(j)}$ and $\beta_+^{(j)}$ may not be adjacent, it is clear that, for large j, the minimum tangent angle of both $\beta_-^{(j)}$ and $\beta_+^{(j)}$ must increase at least ε_0 as time goes from $-M-1$ to $-M$. By Lemma 8.11, the total absolute curvature of $\gamma_j(\cdot, t)$ must drop at least ε_0 during $[-M-1, M]$. By the invariance of the total absolute curvature under scaling, the total curvature of $\gamma(\cdot, t)$ drops at least ε_0 during $[t_j + \varepsilon_j^{1+\sigma}(-M-1), t_j + \varepsilon_j^{1+\sigma}(-M)]$, but this is impossible. So, γ_∞ must be convex and, hence, is embedded.

Our next claim is that the total absolute curvature of γ_∞ must be equal to π. In fact, if it is less than π at some $t_0 \leqslant 0$, we can represent $\gamma_\infty(\cdot, t_0)$ as the graph of some convex function $u(x, t_0)$ over the entire x-axis. By the same reasoning in the proof of Proposition 5.4, the whole family $\gamma(\cdot, t)$ can also be represented as graphs of convex functions $u(x, t)$ whose gradients are uniformly bounded. It follows from the following lemma that $u_{xx} \equiv 0$, contradicting $|k_\infty(0, 0)| = 1$.

Lemma 8.16 Let $u : \mathbb{R} \times [0, \infty) \to [0, \infty)$ be a convex solution of

$$u_t = \frac{u_{xx}^\sigma}{(1 + u_x^2)^{\frac{1}{2}(3\sigma - 1)}}, \tag{8.21}$$

where $|u_x|$ and $|u_{xx}|$ are uniformly bounded. Then,

$$|u_{xx}(x, t)| \leqslant Ct^{-\frac{1}{1+\sigma}}, \quad \forall (x, t) \in [-2, 2] \times [2, \infty).$$

Proof: Let $A = \sup\{|u_x(x, t)| : (x, t) \in \mathbb{R} \times [0, \infty)\}$. Without loss of generality, assume $u(0, 1) = 1$. Let v be an expanding self-similar solution of (8.1) satisfying $v(x) > u(x, 1)$ and $\sup |v_x| = A + 1$ (see

§2.1). By the comparison principle,

$$0 \leqslant u(x,t) \leqslant \frac{1}{1+\sigma} t^{\frac{1}{1+\sigma}} v((1+\sigma)t^{\frac{-1}{1+\sigma}}x) \qquad (8.22)$$

in $\mathbb{R} \times [1,\infty)$.

Let

$$w(x,t) = c\,\varphi\left(\frac{x-\xi}{c}\right) + \left(\frac{1}{c}\right)^{\sigma}(t-1) + \beta$$

be the travelling wave solution of (8.1) with speed c. Recall that φ is an even, nonnegative convex function in $(-\alpha,\alpha)$, $\alpha \geqslant 2$, satisfying $\varphi(0) = 0$ and

$$|\varphi'(x)| > A \quad \text{in } (-\alpha,-2] \cup [2,\alpha). \qquad (8.23)$$

Fix (x_0,t_0) where $x_0 \in [-2,2]$ and $t_0 \geqslant 2$. By (8.22),

$$0 \leqslant u(x_0,t_0) \leqslant Bt_0^{\frac{1}{1+\sigma}}, \qquad (8.24)$$

where $B = \max\{v(x) : x \in [-4,4]\}$. Now we first choose the wave speed $c = (2B)^{-\frac{1}{\sigma}} t_0^{\frac{1}{1+\sigma}}$. By (8.23) and (8.24), we can choose $\xi \in [-2-2c, 2+2c]$ and $\beta < 0$ such that

$$u(x_0,t_0) = w(x_0,t_0)$$

and

$$u_x(x_0,t_0) = w_x(x_0,t_0).$$

On the other hand, it is easy to see that the graphs of u and w intersect exactly twice at $t=1$. By applying the Sturm oscillation theorem to $u-w$, we conclude that $u-w$ has, at most, two zeroes (counting multiplicity) for $t>1$. It follows that

$$u_{xx}(x_0,t_0) \;\leqslant\; \frac{1}{c}\,\varphi_{xx}\left(\frac{x_0-\xi}{c}\right)$$

$$\leqslant\; \frac{(2B)^{\frac{1}{\sigma}} \sup\left\{\varphi_{xx}(x) : x \in [-2-\frac{\bar\alpha-2}{2}, 2+\frac{\bar\alpha-2}{2}]\right\}}{t_0^{\frac{1}{1+\sigma}}}$$

for all $x_0 \in [-2, 2]$ and $t_0 \geqslant 8^{1+\sigma}(2B)^{1+\frac{1}{\sigma}}(\overline{\alpha}-2)^{-(1+\sigma)}$, $\overline{\alpha} = \min\{\alpha, 3\}$.

□

We have shown that γ_∞ always has total absolute curvature π. Consequently, for any $\varepsilon > 0$, we can choose two points on $\gamma_\infty(\cdot, 0)$ such that the ratio of the extrinsic distance d_ε and the intrinsic distance ℓ_ε at these two points is less than ε, i.e.,

$$d_\varepsilon / \ell_\varepsilon < \varepsilon. \tag{8.25}$$

Recall the choice of a and b in (8.19) and (8.20). We consider the extrinsic and intrinsic distances in $[a, b] \times [t_0, \omega)$, where t_0 is chosen so that (8.19) holds in $[t_0, \omega)$. We can find some $\delta_1 > 0$ such that

$$\inf\left\{\frac{d(a, q, t)}{\ell(a, q, t)} : q \in [a, b], \ t \in [t_0, \omega)\right\} \geqslant \delta_1$$

and

$$\inf\left\{\frac{d(p, b, t)}{\ell(p, b, t)} : p \in [a, b], \ t \in [t_0, \omega)\right\} \geqslant \delta_1.$$

By Proposition 8.4, (we choose $\varepsilon < \delta_1$)

$$\inf\left\{\frac{d(p, q, t)}{\ell(p, q, t)} : p, q \in [a, b], \ t \in [t_0, \omega)\right\}$$
$$\geqslant \ \inf\left\{\delta_1, \ \min\left\{\frac{d(p, q, t_0)}{d(p, q, t_0)} : p, q \in [a, b]\right\}\right\}.$$

On the other hand, by (8.25), this is impossible. The contradiction shows that every regular arc of γ^* must be flat. Proposition 8.15 is proved.

Finally, we prove the main result of this section.

Theorem 8.17 *Any maximal solution $\gamma(\cdot, t)$ of (8.1) converges to a point or a line segment in the Hausdorff metric as $t \uparrow \omega$.*

Proof: By what we have shown, we know that, for any small $\varepsilon > 0$, there exists t_0 close to ω such that, for all $t \in [t_0, \omega)$, $\gamma(\cdot, t)$ is contained in the ε^2-neighborhood of γ^* and the total absolute curvature of $\gamma(\cdot, t)$ outside $\bigcup_i D_\varepsilon(Q_i)$ is less than ε. In particular, each component of $\gamma(\cdot, t) \setminus \bigcup_i D_\varepsilon(Q_i)$ is a graph over the corresponding line segment in $\gamma^* \setminus \bigcup_i D_\varepsilon(Q_i)$. By counting the crossing number of $\gamma(\cdot, t) \setminus \bigcup_i D_\varepsilon(Q_i)$ with suitable transversal line segments, we may assume that the number of graphs over a line segment in $\gamma^* \setminus \bigcup_i D_\varepsilon(Q_i)$ is constant for all t.

Suppose on the contrary that γ^* has two line segments Γ_1 and Γ_2 which meet at Q with an angle not equal to zero or π. It is not hard to see that we can find a connected arc $C(t_0)$ of $\gamma(\cdot, t_0)$ satisfying

(i) $C(t_0)$ is contained in $D_{r_0/2}(Q)$, where $r_0 = \min\{|Q_i - Q_j)| : i \neq j\}$, and

(ii) $C(t_0) \setminus D_\varepsilon(Q)$ consists of two components $C_1(t_0)$ and $C_2(t_0)$ so that $C_i(t_0)$ is a graph of some function over $\Gamma_i \cap (D_{r_0/2}(Q) \setminus D_\varepsilon(Q))$ for $i = 1, 2$.

We follow the evolution of $C(t_0)$ to obtain a connected evolving arc $C(t)$ of $\gamma(\cdot, t)$ for all $t \in [t_0, \omega)$ satisfying

(i)' $C(t)$ is contained in $D_{r_0/2}(Q)$ and converges to $\gamma^* \cap D_{r_0/2}(Q)$ in the Hausdorff metric as $t \uparrow \omega$, and

(ii)' $C(t) \setminus D_\varepsilon(Q)$ consists of two components $C_1(t)$ and $C_2(t)$ so that $C_i(t), i = 1, 2$, is the graph over $\Gamma_i \cap (D_{r_0/2}(Q) \setminus D_\varepsilon(Q))$ which converges to $\Gamma_i \cap (D_{r_0/2}(Q) \cap D_\varepsilon(Q))$ in C^2-norm.

Let's examine d/ℓ for two points on $C(t)$. As before, by Proposition 8.4, we have

$$\inf\left\{\frac{d(p, q, t)}{\ell(p, q, t)} : p.q \in C(t), t \in [t_0, \omega)\right\}$$
$$\geqslant \delta_0 > 0$$

for some δ_0. On the other hand, $C(t)$ converges to $\gamma^* \cap D_{r_0/2}(Q)$ and Q is a singularity. By repeating the blow-up argument in the proof of Theorem 8.15, we have a contradiction. Hence, γ^* must be a single line segment or a single point. □

8.4   Removal of interior singularities

In the last section, we have shown that γ^* is a line segment. It may happen that there are some singularities lying on its interior. Now we prepare to show that this is impossible. First of all, we need to establish a whisker lemma for (8.1).

In view of the discussion in previous sections, we may assume the number of convex\concave arcs is constant in time. Furthermore, their total absolute curvature is either greater than π or less than $\pi - \delta_0$ for some positive δ_0. By the definition of convex and concave arcs, we know that each concave arc cannot have interior inflection points, and, for a convex arc, its inflection spots, if they exist, always lie at the ends.

For any real number β, we call $P \in \gamma(\cdot, t)$ a β-**point** if its tangent angle is β (mod 2π).

Lemma 8.18 *Let A_1 be a β-point on $\gamma(\cdot, t)$ for some $t_1 \in (0, \omega)$. Suppose it is not an inflection point. Then, there exists a C^2-family $\{A_t\}$, $t \in [0, t_1]$, of β-points on $\gamma(\cdot, t)$ such that $A_{t_1} = A_1$.*

Proof: As A_1 is not an inflection point, the implicit function theorem ensures that there exists a C^2-family of β-points for t near $t_1, t \leqslant t_1$. Let $(t_0, t_1]$ be the largest interval on which the family is defined. The point A_t will accumulate at inflection spots of $\gamma(\cdot, t_0)$. However, as any inflection spot is connected by a convex arc and a concave arc, it follows from Lemma 8.6 that there are no β-points around the accumulation point for t slightly above t_0, and contradiction holds. $\square$

An arc $b(t) \subseteq \gamma(\cdot, t)$ is called a β-**inward arc** (resp. β-**outward arc**) if the following hold:

(a) the arc $b(t)$ has negative (resp. positive) curvature,

(b) the inward (resp. outward) pointing unit tangents to $b(t)$ at the endpoints are both given by $(\cos\beta, \sin\beta)$, and

(c) both endpoints are not inflection points of $\gamma(\cdot, t)$.

By this definition, the endpoints of a β-inward arc (resp. β-outward arc) are a β-point and a $(\beta + \pi)$-point, respectively. We can use Lemma 8.18 to trace back in time to form a continuous family of β-inward arcs (resp. β-outward arcs). Now we have the following lemma, whose proof is similar to that of the whisker lemma in §6.4.

Lemma 8.19 (whisker lemma) *There exists $\delta > 0$ depending only on γ_0 such that, for any P on a β-inward arc (or β-outward arc), $b(\bar{t}) \subset \gamma(\cdot, \bar{t})$ with $\bar{t} \in (0, \omega)$, the δ-whisker,*

$$\ell_{P,\delta,\beta} = \{ P + r(\cos\beta, \sin\beta) : \ 0 \leqslant r \leqslant \delta \},$$

is disjoint from $\gamma(\cdot, \bar{t}) \setminus b(\bar{t})$.

The following lemma tells us that any convex or concave arc of total absolute curvature less than $\pi - \delta_0$ disappears as $t \uparrow \omega$.

Lemma 8.20 *There exists $K > 0$ depending only on γ_0 such that the curvature on any convex\concave arc with total absolute curvature less than $\pi - \delta_0$ is bounded between $-K$ and K.*

Proof: Let $C(t)$ be an evolving convex\concave arc with total absolute curvature less than $\pi - \delta_0$. By disposing the inflection spots, we may assume $C(t)$ is free of inflection points and can be parametrized by its tangent angle θ. Then, $k = k(\theta, t)$ satisfies

$$k_t = k^2(v_{\theta\theta} + v), \tag{8.26}$$

where $v = |k|^{\sigma-1} k(\theta, t)$. Without loss of generality, we may assume the tangent range of $C(0)$ is contained in $[\delta_0/2, \pi - \delta_0/2]$. By Lemma

8.11, the tangent range of $C(t)$ lies inside the same interval for all $t > 0$. Equation (8.26) admits two stationary solutions given by

$$k^{\pm} = \pm (A(\sin \frac{\delta_0}{2})^{-1} \sin \theta)^{\frac{1}{\sigma}}, \quad \theta \in [0, 2\pi],$$

where $A = \max_{\gamma_0} |k(\cdot, 0)|^{\sigma}$. It follows from the comparison principle that

$$|k(\theta, t)| \leqslant A^{\frac{1}{\sigma}} \left(\sin \frac{\delta_0}{2} \right)^{-\frac{1}{\sigma}}$$

for all $t \in [0, \omega)$.   □

Now, suppose that γ^* is a line segment $[0, \ell]$ on the x-axis. Let $(0,0) = Q_1 < Q_2 < \cdots < Q_m = (\ell, 0)$ be its singularities. For any small $2\varepsilon < r_0 = \min\{\text{dist}\,(Q_i, Q_j) : i \neq j\}$, there exists t_ε such that, for all $t \geqslant t_\varepsilon$, $\gamma(\cdot, t)$ is contained in the ε^2-neighborhood of γ^*. Furthermore, $\gamma(\cdot, t)$ converges in C^2-norm to γ^* away from $\bigcup_{j=1}^{m} D_\varepsilon(Q_j)$, and the number of connected arcs, which are now graphs of C^2-functions over the intervals of $[0, \ell] \setminus \bigcup_{j=1}^{m} D_\varepsilon(Q_j)$, are constant in time. We are going to show that $m = 2$ by assuming there is an interior singularity Q_2, and then draw a contradiction. Imagine at some time close to ω, an arc enters $D_\varepsilon(Q_2)$ from the side of Q_1 or Q_3. It would stay and wriggle inside $D_\varepsilon(Q_2)$ and then come out either along the same side or the opposite side. According to the different possibilities, we classify the arc into two types.

Specifically, a connected arc $c(t_0)$ of $\gamma(\cdot, t_0)$ is of **type I** if its total absolute curvature is greater than $\pi - \delta_0/2$ and it connects a point on the vertical line $\{x = Q_2 - r_0/2\}$ to a point on the vertical line $\{x = Q_2 + r_0/2\}$ such that $c(t_0) \cap ([Q_2 - r_0/2, Q_2 - r_0/4] \times \mathbb{R})$ and $c(t_0) \cap ([Q_2 + r_0/4, Q_2 + r_0/2] \times \mathbb{R})$ are connected. A connected arc $c(t_0)$ is of **type II** if it connects a vertical point (i.e., β-point with

$\beta = \pm\pi/2$) $P(t_0)$ to a point $R(t_0)$ on the vertical line $\{x = Q_2 + r_0/2\}$ (or $\{x = Q_2 - r_0/2\}$) such that the following hold:

(II$_1$) $c(t_0) \cap ([Q_2 + r_0/4, Q_2 + r_0/2] \times \mathbb{R})$ (or $c(t_0) \cap ([Q_2 - r_0/2, Q - 2 - r_0/4] \times \mathbb{R}))$ is connected,

(II$_2$) $c(t_0) \cap ([Q_2 - r_0/2, Q_2 - r_0/4] \times \mathbb{R})$ (or $c(t_0) \cap ([Q_2 + r_0/4, Q_2 + r_0/2] \times \mathbb{R})$ has exactly two components $b_1(t_0)$ and $b_2(t_0)$. Here, $b_1(t_0)$ is the subarc which is closer to $P(t_0)$ and $b_2(t_0)$ is closer $R(t_0)$,

(II$_3$) $P(t_0)$ is not an inflection point, and it lies on a convex\concave arc $\Gamma(t_0)$ whose total curvature is greater than π,

(II$_4$) the total absolute curvature of the subarc $d_1(t_0)$ connecting $P(t_0)$ and $b_1(t_0)$ is nearly $\pi/2$,

(II$_5$) the total absolute curvature of the subarc $d_2(t_0)$ connecting $R(t_0)$ and $b_2(t_0)$ is nearly 0 , and

(II$_6$) denoting the tangent vector of $c(t_0)$ at $P(t_0)$ along the direction from $P(t_0)$ to $R(t_0)$ by $\boldsymbol{e}$, the ray $\{r\boldsymbol{e} : r > 0\}$ intersects to $d_2(t_0)$.

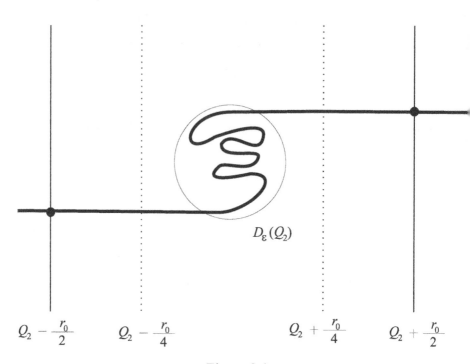

$$Q_2 - \frac{r_0}{2} \qquad Q_2 - \frac{r_0}{4} \qquad\qquad\qquad Q_2 + \frac{r_0}{4} \qquad Q_2 + \frac{r_0}{2}$$

Figure 8.1
A type I-arc near Q_2

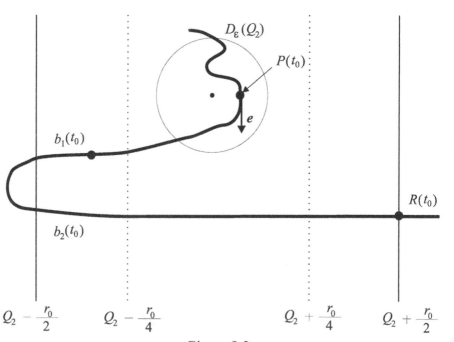

$$Q_2 - \frac{r_0}{2} \qquad Q_2 - \frac{r_0}{4} \qquad\qquad\qquad Q_2 + \frac{r_0}{4} \qquad Q_2 + \frac{r_0}{2}$$

Figure 8.2
A type II-arc near Q_2

Lemma 8.21 *For any small $\varepsilon > 0$, there exists some t_0 close to ω such that $\gamma(\cdot, t_0)$ admits either a type I or a type II arc around an interior singularity.*

Proof: Let $\Gamma(t_0)$ be an evolving convex\convex arc developing the singularity Q_2, and $O(t_0)$ a vertical point on $\Gamma(t_0)$. We shall find a type I\type II arc by tracing $\gamma(\cdot, t_0)$ starting at $O(t_0)$.

Let's first trace along the curve from $O(t_0)$ in the counterclockwise direction. It is obvious that we can find a connected arc $c_1'(t_0)$ which connects $O(t_0)$ to a point $G_1(t_0)$ on the vertical line $\{x = Q_2 - r_0/2\}$ (or $\{x = Q_2 + r_0/2\}$) such that $c_1'(t_0) \cap ([Q_2 - r_0/2, Q_2 - r_2/4] \times \mathbb{R})$ (or $c_1'(t_0) \cap ([Q_2 + r_0/4, Q_2 + r_0/2] \times \mathbb{R})$) is connected and $c_1'(t_0) \cap ([Q_2 + r_0/4, Q_2 + r_0/2] \times \mathbb{R}) = \phi$ (or $c_1'(t_0) \cap ([Q_2 - r_0/2, Q_2 - r_0/4] \times \mathbb{R} = \phi)$. By the same reasoning, along the clockwise direction, we can find $c_2'(t_0)$ connecting $O(t_0)$ and some point $G_2(t)$ lying on $\{x = Q_2 - r_0/2\}$ (or $\{x = Q_2 + r_0/2\}$) such that $c_2'(t_0) \cap ([Q_2 - r_0/2, Q_2 - r_0/4] \times \mathbb{R})$ (or $c_2'(t_0) \cap ([Q_2 + r_0/4, Q_2 + r_0/2] \times \mathbb{R})$) is connected and $c_2'(t_0) \cap ([Q_2 + r_0/4, Q_2 + r_0/2] \times \mathbb{R}) = \phi$ (or $c_2'(t_0) \cap ([Q_2 - r_0/2, Q_2 - r_0/4] \times \mathbb{R}) = \phi)$. If $G_1(t_0)$ and $G_2(t_0)$ lie on different vertical lines, the union of $c_1'(t_0)$ and $c_2'(t_0)$ forms a type I arc. If not, we consider Case (1) $G_1(t_0)$ and $G_2(t_0)$ lie on $\{x = Q_2 - r_0/2\}$, and Case (2) $G(t_0)$ lies on $\{x = Q_2 + r_0/2\}$ separately.

Consider Case (1). When tracing the curve in the counterclockwise direction from $G_1(t_0)$, it is clear that we can find an arc $c_1''(t_0)$ connecting $G_1(t_0)$ to a point $R(t_0)$ on the vertical line $\{x = Q_2 + r_0/2\}$ such that $c_1''(t_0) \cap ([Q_2 + r_0/4, Q_2 + r_0/2] \times \mathbb{R})$ is connected. Now we want to choose a better vertical point on $c_1'(t_0) \cup c_1''(t_0)$ to replace $O(t_0)$. We start at $R(t_0)$ and trace back along $c_1'(t_0) \cup c_1''(t_0)$. Since $\gamma(\cdot, t_0)$ lies inside the ε^2-neighborhood of γ^*, we will go in and out of the disk $D_\varepsilon(Q_2)$. Then it goes into $D_\varepsilon(Q_1)$ and then returns to hit a point $P'(t_0)$ on $\partial D_\varepsilon(Q_2)$. Since $c_1'(t_0)$ and $c_1''(t_0)$ are connected in $[Q_2 + r_0/4, \, Q_2 + r_0/2] \times \mathbb{R}$, the curve cannot leave the disk to the right. This means that, as tracing from the first re-hit point $P'(t_0)$ along $c_1'(t_0) \cup c_1''(t_0)$, we arrive at a vertical point, $P(t_0)$, for the first time. In the following, we show that the subarc, denoted by $c(t_0)$, of $c_1'(t_0) \cup c_1''(t_0)$, which connects $P(t_0)$ to $R(t_0)$, is of type II.

First of all, it follows from the definition of $c(t_0)$ that (II_1) and (II_2) are fulfilled. The sum of the total absolute curvature of the "small" convex\concave arc is arbitrarily small when t_0 is sufficiently close to ω. As we trace from $P'(t_0)$ to $P(t_0)$, the tangent turns nearly $\pi/2$, and so it must meet a "large" convex\concave $\Gamma(t_0)$. Together with the fact that $P(t_0)$ is the first vertical point from $P'(t_0)$, we know that (II_3) and (II_4) also hold. Now, consider the subarc $d_2(t_0)$ connecting $R(t_0)$ to $b_2(t_0)$. Suppose that the total absolute curvature of $d_2(t_0)$ is not arbitrarily small. Then, $d_2(t_0)$ contains a convex\concave subarc whose total curvature inside $D_\varepsilon(Q_2)$ is not less than $\pi - \delta_0/4$. This says that $d_2(t_0)$ is of type I. So, we may always assume that (II_5) holds in this case. Finally, to verify (II_6), we trace the curve from $b_1(t_0)$ into $D_\varepsilon(Q_1)$ in counterclockwise direction. Since Q_1 is the left endpoint of γ^*, we will stop at a first vertical point, $P''(t_0)$, in $D_\varepsilon(Q_1)$. The same reasoning as above shows that $P''(t_0)$ is not an inflection point, and it lies on a convex\concave arc $\Gamma''(t_0)$ whose total curvature is greater than π, and the total absolute curvature of the arc $d_3(t_0)$ of $c(t_0)$, which connects $b_1(t_0)$ to $P''(t_0)$, is nearly $\pi/2$. It follows from the whisker lemma that $\Gamma''(t_0)$ is a convex arc, and the δ-whisker at $P''(t_0)$, $\ell_{P''(t_0),\delta,\theta}$ (θ is close to π), is disjoint from $\gamma(\cdot, t_0) \backslash \{P''(t_0)\}$. Also, the δ-whisker at $P(t_0)$, $\ell_{P(t_0),\delta,0}$, is disjoint from $\gamma(\cdot, t_0) \backslash \{P(t_0)\}$. If the arc $\Gamma(t_0)$ which contains $P(t_0)$ is convex, any point slightly passing beyond $P(t_0)$ in the clockwise direction would not be able to be connected to $G_2(t_0)$ by avoiding the δ-whisker at $P(t_0)$. Thus, $\Gamma(t_0)$ must be concave. Furthermore, the δ-whisker at $P''(t_0)$ forces (II$_6$) to hold in this case.

Next, we consider Case (2). Exactly as before, we trace along the curve starting at $G_1(t_0)$ in the counterclockwise direction to get an subarc $c_1''(t_0)$ connecting $G_1(t_0)$ to some $R_1(t_0)$ on $\{x = Q_2 - r_0/2\}$ so that $c_1''(t_0) \cap ([Q_2 - r_2/2, Q_2 - r_0/4] \times \mathbb{R})$ is connected. Then, by tracking back from $R_1(t_0)$ along $c_1'(t_0) \cup c_1''(t_0)$, we find a vertical point $P_1(t_0)$ on $c_1'(t_0) \cup c_1''(t_0)$ and a sub-arc $c_1(t_0)$ of $c_1'(t_0) \cup c_1''(t_0)$ connecting $P_1(t_0)$ and $R_1(t_0)$ such that (II$_1$) – (II$_5$) hold with, $c(t_0), P(t_0), R(t_0), b_1(t_0), b_2(t_0), d_1(t_0), d_2(t_0)$ replaced by $c_1(t_0), P_1(t_0), R_1(t_0), b_1^{(1)}(t_0), b_2^{(1)}(t_0), d_1^{(1)}(t_0), d_2^{(1)}(t_0)$, accordingly. Similarly, we trace the curve starting at $G_2(t_0)$ along the clockwise direction to get a connected subarc $c_2''(t_0)$ connecting $G_2(t_0)$ to some $R_2(t_0)$ on the vertical line $\{x = Q_2 - r_0/2\}$. Then, we trace back from $R_2(t_0)$ along $c_2'(t_0) \cup c_2''(t_0)$ to get a subarc $c_2(t_0)$ of $c_2'(t_0) \cup c_2''(t_0)$ which connects a vertical point $P_2(t_0)$ to $R_1(t_0)$ such that (II$_1$) – (II$_5$) hold with $c(t_0), P(t_0) \cdots$, replaced by $c_2(t_0), P_2(t_0) \cdots$, accordingly.

The same as in Case (1), it follows from (II$_3$), (II$_4$), and the whisker lemma that the δ-whisker at $P_1(t_0)$, $\ell_{P_1(t_0),\delta,\pi}$, is disjoint from $\gamma(\cdot, t_0) \setminus \{P_1(t_0)\}$ and the δ-whisker at $P_2(t_0)$, $\ell_{P_2(t_0),\delta,\pi}$, is disjoint from $\gamma(\cdot, t_0) \setminus \{P_2(t_0)\}$. Also, as we continue to trace along the curve from $R_1(t_0)$ to $R_2(t_0)$ to the left, we will arrive at a corresponding first vertical point $P_1''(t_0)$ and $P_2''(t_0)$ in $D_\varepsilon(Q_1)$ such that the δ-whisker at $P_1''(t_0)$, $\ell_{P_1''(t_0),\delta,\theta_1}$, where θ_1 is close to π, is disjoint from $\gamma(\cdot, t_0) \setminus \{P_1''(t_0)\}$ and the δ-whisker at $P_2''(t_0)$, $\ell_{P_2''(t_0),\delta,\theta_2}$, where θ_2 close to π, is disjoint from $\gamma(\cdot, t_0) \setminus \{P_2''(t_0)\}$. Let's denote by e_1 (or e_2) the tangent of $c_1(t_0)$ (resp. $c_2(t_0)$) at $P_1(t_0)$ (resp. $P_2(t_0)$) along the direction from $P_1(t_0)$ (resp. $P_2(t_0)$) to $R_1(t_0)$ (resp. $R_2(t_0)$). We claim that either $c_1(t_0)$ or $c_2(t_0)$ satisfies (II$_6$)

For, suppose $c_1(t_0)$ does not satisfy (II$_6$). In other words, the

ray $\{re_1 : r > 0\}$ does not intersect $d_2^{(1)}(t_0)$. Note that $P_1''(t_0)$ lies
on a convex arc. We have already seen that $b_1^{(1)}(t_0), b_2^{(1)}(t_0), b_1^{(2)}(t_0)$,
and $b_2^{(2)}(t_0)$ are disjoint graphs over $[Q_2 + r_0/4, Q_2 + r_0/2]$. The embeddedness of the curve and the δ-whiskers force $b_1^{(2)}(t_0)$ and $b_2^{(2)}(t_0)$
to lie between $b_1^{(1)}(t_0)$ and $b_2^{(1)}(t_0)$, and, hence, $c_2(t_0)$ must satisfy
(II_6). $\square$

Now, we can show:

Proposition 8.22 *There are no interior singularities on γ^*.*

Proof: By Lemma 8.21, it suffices to show that there are no type I\
type II arcs on $\gamma(\cdot, t_0)$ when t_0 is close to ω.

Suppose that $c(t_0)$ is a type I arc. We follow the evolution of
$c(t_0)$ to obtain an evolving arc $c(t)$ of $\gamma(\cdot, t)$, $t \in [t_0, \omega)$, where $c(t)$
connects a point $P(t)$ on $\{x = Q_2 - r_0/2\}$ to another point $R(t)$ on
$\{x = Q_2 + r_0/2\}$ such that both $c(t) \cap ([Q_1 - r_0/2, Q_2 - r_0/4] \times \mathbb{R})$
and $c(t) \cap ([Q_2 + r_0/4, Q_2 + r_0/2] \times \mathbb{R})$ are connected and collapse
to the x-axis in C^2-norm. There is a convex\concave subarc of $c(t_0)$
whose total curvature inside $D_\varepsilon(Q_2)$ is nearly π. (Notice that the
whisker lemma forces that its total curvature cannot be much away
from π.) So $c(t)$ continues to have total absolute curvature close to
π in $D_\varepsilon(Q_2)$ for all $t \in [t_0, \omega)$ as it develops a singularity at Q_2.

We look at the ratio of extrinsic and intrinsic distances of two
points on $c(t)$. Exactly as before, by Proposition 8.4 and the C^2-
convergence of the flow away from the singularities, we know that

$$\inf \left\{ \frac{d(P, Q, t)}{\ell(P, Q, t)} : P, Q \in c(t), t \in [t_0, \omega) \right\} \geqslant \delta_0 > 0$$

for some δ_0. However, as $c(t)$ develops a singularity at Q_2, by repeating the blow-up argument in Proposition 8.15, we deduce that

$$\inf \left\{ \frac{d(P, Q, t)}{\ell(P, Q, t)} : P, Q \in c(t), t \in [t_0, \omega) \right\} = 0,$$

and the contradiction holds. Hence, $\gamma(\cdot, t_0)$ cannot admit any type I arc.

Next, suppose $c(t_0)$ is a type II arc of $\gamma(\cdot, t)$. Without loss of generality, we may assume $c(t_0)$ connects a vertical point $P(t_0)$ in $D_\varepsilon(Q_2)$ to some $R(t_0)$ on $\{x = Q_2 + r_0/2\}$ satisfying (II_1) – (II_6). Recall that $P(t_0)$ is a non-inflection point on a convex\concave arc $\Gamma(t_0)$ whose total curvature is greater than π. Let $\Gamma(t)$ be the corresponding evolving convex\concave arc from $\Gamma(t_0)$. By the whisker lemma, the total curvature of $\Gamma(t)$ tends to π and the unique vertical point $P(t)$ is always a non-inflection point. By the implicit function theorem, $P(t)$ is C^2 in $t \in [t_0, \omega)$. We can follow the evolution of $c(t_0)$ to get a connected evolving arc $c(t)$ of $\gamma(\cdot, t)$ which connects $P(t)$ to some $R(t)$ on $\{x = Q_2 + r_0/2\}$ such that for all $t \in [t_0, \omega)$,

$(\mathrm{II}_1)'$ $c(t) \cap ([Q_2 + r_2/4, Q_2 + r_0/2] \times \mathbb{R})$ is connected,

$(\mathrm{II}_2)'$ $c(t) \cap ([Q_2 - r_0/2, Q_2 - r_0/4] \times \mathbb{R})$ has exactly two components $b_1(t)$ and $b_2(t)$, where $b_1(t)$ is the arc closer to $P(t)$,

$(\mathrm{II}_3)'$ $P(t)$ is the only vertical point of $\Gamma(t)$,

$(\mathrm{II}_4)'$ the total absolute curvature of the subarc $d_1(t)$ of $c(t)$ which connects $P(t)$ and $b_1(t)$ tends to $\pi/2$,

$(\mathrm{II}_5)'$ the total absolute curvature of the subarc $d_2(t)$ of $c(t)$ which connects $R(t)$ and $b_2(t)$ tends to 0, and

$(\mathrm{II}_6)'$ the vertical vector $\boldsymbol{e}$, which is the tangent at $P(t)$ along the direction from $P(t)$ to $R(t)$, points in such a way that the ray $\{r\boldsymbol{e} : r > 0\}$ intersects $d_2(t)$.

Here, in $(\mathrm{II}_4)'$ and $(\mathrm{II}_5)'$, we have used C^2-convergence of the flow away from the singularities and the fact that the spherical image of each evolving convex\concave arc is strictly nesting in time.

As before, consider the ratio of extrinsic and intrinsic distances on $c(t)$. We claim that, for each $t \in [t_0, \omega)$, the ratio attains its

minimum in the interior of $c(t) \times c(t)$. For, the boundary of $c(t) \times c(t)$
consists of $\{P(t)\} \times c(t)$ and $c(t) \times \{R(t)\}$. By the C^2-convergence
of $c(t)$ inside $[Q_2 + r_0/4, Q_2 + r_0/2] \times \mathbb{R}$, it is clear that the minimum
of d/ℓ over $\{P(t)\} \times c(t)$ is strictly less than the its minimum over
$c(t) \times \{R(t)\}$ when t_0 is close to ω. It is also clear that there is some
interior point $P'(t)$ on $c(t)$ such that $d(P(t), P'(t), t)/\ell(P(t), P'(t), t)$
attains the minimum of d/ℓ inside $\{P(t)\} \times c(t)$. For simplicity, let's
assume the arc length parametrization satisfies $s(P'(t)) > s(P(t))$.
Let

$$\omega = \frac{\gamma(P(t), t) - \gamma(P'(t), t)}{|\gamma(P(t), t) - \gamma(P'(t), t)|}$$

and

$$e' = \frac{d}{ds} \gamma(P'(t), t).$$

At $(P(t), P'(t), t)$, we have

$$0 = \frac{d}{ds}\Big|_{s=0} \frac{|\gamma(s(P'(t)) + s, t) - \gamma(s(P(t)), t)|}{\ell + s}$$

$$= \frac{1}{\ell}\langle -\omega, e' \rangle - \frac{d}{\ell^2}.$$

So,

$$\langle \omega, e' \rangle = -\frac{d}{\ell}. \tag{8.27}$$

By definition, the right-hand side of (8.27) can be arbitrarily small
as t_0 tends to ω. It follows from (II'_4), $(II_5)'$, and (8.27) that $P'(t)$
lies on the subarc $d_2(t)$, and ω approaches $-e$ as $t \uparrow \omega$. We compute

$$\frac{d}{ds}\Big|_{s=0} \left(\frac{|\gamma(s(P(t)) + s, t) - \gamma(s(P'(t)), t)|}{\ell - s} \right)$$

$$= \frac{1}{\ell}(\langle \omega, e \rangle + \frac{d}{\ell}) < 0,$$

as $t_0 \uparrow \omega$. Thus, we can find some interior point $P''(t)$ with arc length parameter $s(P(t)) + s_0$ $(s_0 > 0)$ such that

$$\frac{d(P''(t), P'(t), t)}{\ell(P''(t), P'(t), t)} < \frac{d(P(t), P'(t), t)}{\ell(P(t), P'(t), t)}.$$

So, d/ℓ attains its minimum in the interior. By Proposition 8.4,

$$\inf \left\{ \frac{d(P, Q, t)}{\ell(P, Q, t)} : P, Q \in c(t), t \in [t_0, \omega) \right\}$$
$$\geqslant \quad \delta_0 > 0$$

for some positive δ_0. On the other hand, as $c(t)$ develops a singularity at some Q_j with $j \neq 2$, this infimum should be zero. This leads to a contradiction. The proof of Proposition 8.22 is completed. □

8.5 The almost convexity theorem

Finally, we can prove the Theorem 8.1.

In view of Theorem 8.17 and Proposition 8.22, it remains to show that there is no evolving concave arc with total curvature greater than π at the endpoints for all t sufficiently close to ω. Suppose that there is such an evolving concave arc $\Gamma(t)$ which persists at the end. We can find a β-inward arc $b(t) \subseteq \Gamma(t)$ $(\beta = 0$ or $\pi)$, and so γ^* cannot be a point. So, for all small $\varepsilon > 0$, $\gamma(\cdot, t) \setminus D_\varepsilon(Q_1) \cup D_\varepsilon(Q_2)$ consists of exactly two arcs converging to a line segment in C^2-norm. Without loss of generality, we assume $\Gamma(t)$ develops a singularity at Q_1.

Let $G_1(t)$ and $G_2(t)$ be two points of $\gamma(\cdot, t)$ lying on the middle line $\{x = \ell/2\}$ and $G_1(t)$ lies below $G_2(t)$. To understand the behaviour of $\gamma(\cdot, t)$ inside $D_\varepsilon(Q_1)$, we trace the curve from $G_1(t)$ in the clockwise direction. It goes into $D_\varepsilon(Q_1)$ and hits a first vertical point $P_1(t)$. Since the number of convex\concave arcs is fixed and their total curvature either tends to 0 or is always greater than π as

$t \uparrow \omega$, $P_1(t)$ must be located on a convex\concave arc $\Gamma_1(t)$ whose total curvature is greater than π, and the total absolute curvature of the arc connecting $G_1(t)$ and $P_1(t)$ tends to $\pi/2$. By the whisker lemma, $\Gamma_1(t)$ must be convex. We claim that the total curvature of $\Gamma_1(t)$ tends to π.

To see this, we trace the curve from $\Gamma_1(t)$ along the clockwise direction. After passing through one or more "small convex\concave arcs," we will arrive at a concave\convex arc $\Gamma_2(t)$ whose total curvature is greater then π. If the total curvature of $\Gamma_1(t)$ is uniformly greater than π, there is some β-inward arc (or β-outward arc) $b(t)$ on the second "large arc" $\Gamma_2(t)$ such that the corresponding δ-whisker on $b(t)$ crosses $\gamma(\cdot, t) \setminus b(t)$. This contradicts the whisker lemma. Hence, the total curvature of $\Gamma_1(t)$ tends to π at the end.

Let's examine the next "large arc," $\Gamma_2(t)$. The total absolute curvature of the arc between $\Gamma_1(t)$ and $\Gamma_2(t)$ tends to zero. There exists a vertical point $P_2(t)$ on $\Gamma_2(t)$. Again, by definition and the whisker lemma, there is a β-inward arc (or β-outward arc) on $\Gamma_2(t)$ with $\beta = 0$, and the total curvature of $\Gamma_2(t)$ also tends to π. Observe that $\Gamma_2(t)$ is concave. In fact, if it is convex, the horizontal δ-whisker $\ell_{P_2(t),\delta,0}$ would prevent the points slightly beyond $P_2(t)$ in the clockwise direction from connecting $G_2(t)$ along $\gamma(\cdot, t)$. Continue to trace along $\gamma(\cdot, t)$ in the clockwise direction. After passing through some "small arcs" from $\Gamma_2(t)$, we will get into the third "large arc," $\Gamma_3(t)$, with a vertical point $P_3(t)$. It is clear that $\Gamma_3(t)$ is convex. If we do not meet any "large arc" by tracing from $\Gamma_3(t)$ to $G_2(t)$, the total absolute curvature of the arc between $\Gamma_3(t)$ and $G_2(t)$ is very small. So, the arc between $\Gamma_3(t)$ and $G_2(t)$ is nearly horizontal. The total curvature of $\Gamma_3(t)$ also tends to π, and the total absolute curvature of the arc connecting $P_3(t)$ and $G_2(t)$ tends to $\pi/2$. If we meet a fourth "large arc," $\Gamma_4(t)$, then, exactly as before, the total curvature of $\Gamma_3(t)$ tends to π, and $\Gamma_4(t)$ is concave with total curvature nearly equal to π, too. By repeating this argument, we can decompose the arc from $G_1(t)$ to $G_2(t)$ into "large" and "small arcs" where the "large arcs," $\Gamma_1(t), \Gamma_2(t), \cdots, \Gamma_{2n-1}(t), (n \geqslant 2)$, satisfy (i) the total curvature of each $\Gamma_i(t)$ tends to π, and (ii) $\Gamma_{2k-1}(t)$ is convex and $\Gamma_{2k}(t)$ is concave. For each $i, 1 \leqslant i \leqslant 2n - 1$, there is a unique path $P_i(t)$ consisting of vertical points on $\Gamma_i(t)$ which is C^2 for $t \in [t_0, \omega)$.

Consider $\gamma(\cdot, t)$ inside $D_\varepsilon(Q_1)$. Let $R(t) \in \gamma(\cdot, t) \cap D_\varepsilon(Q_1)$ be the maximum point of $|k(\cdot, t)|$ on $\gamma(\cdot, t) \cap D_\varepsilon(Q_1)$. There exists a sequence $\{t_j\}$ which converges to ω such that

$$|k(\cdot, t)| \leqslant |k(R(t_j), t_j)|, \quad \forall\, t \leqslant t_j ,$$

on $\gamma(\cdot, t) \cap D_\varepsilon(Q_1)$. By Lemma 8.20, $R(t_j)$ lies on some "large arc." Without loss of generality, we may assume all $R(t_j)$ lie on a single $\Gamma_{i_0}(t_j), i_0 \in [1, 2n - 1]$. By using a blow-up argument, we conclude as before that

$$\inf \left\{ \frac{d(P, Q, t)}{\ell(P, Q, t)} : P, Q \in \Gamma_{i_0}(t), t \in [t_0, \omega) \right\} = 0. \tag{8.28}$$

Now we consider two cases (a) $i_0 = 1$ or $2n - 1$, and (b) $i < i_0 < 2n - 1$ separately. Take $i_0 = 1$ in Case (a). Now we are in a similar situation of a type II arc near an interior singularity. Denote by $c_1(t)$ the subarc connecting $G_1(t)$ to the vertical point $P_2(t)$ on $\Gamma_2(t)$ along the clockwise direction. As before, consider the ratio d/ℓ on $c_1(t) \times c_1(t)$. It is clear that the minimum of d/ℓ restricted on the boundary of $c_1(t) \times c_1(t)$ is attained at $P_2(t)$ and some interior point $P'(t) \in c_1(t)$. In the following, we assume the arc length parametrization is along the counterclockwise direction. Set

$$\omega_1 = \frac{\gamma(P_2(t), t) - \gamma(P'(t), t)}{|\gamma(P_2(t), t) - \gamma(P'(t), t)|} ,$$

$$\boldsymbol{e}_2 \;\; = \;\; \frac{d}{ds}\gamma(P_2(t), t) \qquad \text{and}$$

$$\boldsymbol{e}' \;\; = \;\; \frac{d}{ds}\gamma(P'(t), t).$$

We have

$$0 \;\; = \;\; \frac{d}{ds}\Big|_{s=0} \left(\frac{|\gamma(s(P'(t)) + s, t) - \gamma(s(P_2(t)), t)|}{\ell + s} \right)$$

$$= \;\; -\frac{1}{\ell}\left(\langle \boldsymbol{\omega}_1, \boldsymbol{e}' \rangle + \frac{d}{\ell} \right), \tag{8.29}$$

$$\frac{d}{ds}\Big|_{s=0} \left(\frac{|\gamma(s(P_2(t)) + s, t) - \gamma(s(P'(t)), t)|}{\ell - s} \right)$$

$$= \;\; \frac{1}{\ell}\left(\langle \boldsymbol{\omega}_1, \boldsymbol{e}_2 \rangle + \frac{d}{\ell} \right). \tag{8.30}$$

In the following, we want to show that there exists a positive ε_0 such that, whenever the minimum of d/ℓ over $c_1(t) \times c_1(t)$ is less than ε_0, d/ℓ, it attains its minimum in the interior of $c_1(t) \times c_1(t)$.

Let's assume that the minimum of d/ℓ over $c_1(t) \times c_1(t)$ is equal to the minimum over the boundary. By an algebraic argument, we may also assume the line segment between $P_2(t)$ and $P'(t)$ has no interior intersection with $c_1(t)$. In fact, suppose then is an interior intersection $\widetilde{P}(t)$. We let d_1 and ℓ_1 be the extrinsic and intrinsic distances between $P_2(t)$ and $\widetilde{P}(t)$, and let d_2 and ℓ_2 be the extrinsic and intrinsic distances between $\widetilde{P}(t)$ and $P'(t)$. Then $d = d_1 + d_2$ and $\ell = \ell_1 + \ell_2$. By definition,

$$\frac{d_1 + d_2}{\ell_1 + \ell_2} \leqslant \frac{d_1}{\ell_1}.$$

We have

$$\frac{d_2}{\ell_2} \leqslant \frac{d_1}{\ell_1},$$

and then

$$\frac{d_2}{\ell_2} \leqslant \frac{d_1 + d_2}{\ell_1 + \ell_2},$$

which shows that the value of d/ℓ at $(\tilde{P}(t), P'(t))$ is not greater than the minimum of d/ℓ over the boundary. Hence, we may always assume the interior of the line segment does not touch $c_1(t)$.

When the minimum of d/ℓ over $c_1(t) \times c_1(t)$ is less than a very small ε_0, (8.29) shows that $\boldsymbol{\omega}_1$ is nearly orthonormal to $\boldsymbol{e}'$. It is easy to see that $P'(t)$ cannot lie on the arc between $P_1(t)$ and $P_2(t)$. To avoid interior intersection, $\boldsymbol{e}'$ is nearly horizontal, and then $\boldsymbol{\omega}_1$ is close to $-\boldsymbol{e}_2$. Thus, by (8.30), d/ℓ attains its minimum in the interior of $c_1(t) \times c_1(t)$.

Now we can apply Proposition 8.4 to conclude that d/ℓ has a positive lower bound in $c_1(t) \times c_1(t)$ for all $t \in [t_0, \omega)$, a contradiction to (8.28).

Next, we treat Case (b). Assume that $i_0 = 2$. We let $c_2(t)$ be the subarc connecting $P_1(t)$ to $P_3(t)$ in the clockwise direction. It is clear that the minimum of d/ℓ over $\partial(c_2(t) \times c_2(t))$ is attained at some $(P''(t), P_3(t))$, $P''(t) \in c_2(t)$. Set

$$\boldsymbol{\omega}_2 = \frac{\gamma(P_3(t), t) - \gamma(P''(t), t)}{|\gamma(P_3(t), t) - \gamma(P''(t), t)|},$$

$$\boldsymbol{e}_1 = \frac{d}{ds}\gamma(P_1(t), t),$$

$$\boldsymbol{e}_3 = \frac{d}{ds}\gamma(P_3(t), t), \qquad \text{and}$$

$$\boldsymbol{e}'' = \frac{d}{ds}\gamma(P''(t), t),$$

where the arc length parametrization is along the counterclockwise direction.

We first claim that $P''(t)$ belongs to the interior of $c_2(t)$. In fact, if $P''(t)$ is an endpoint, then it must be $P_1(t)$. The algebraic argument in the last paragraph shows that the line segment between $P_1(t)$ and $P_3(t)$ has no interior intersection with $c_2(t)$. Because $P_1(t)$ and $P_3(t)$ are vertical points, the line segment is vertical. However, by minimality,

$$0 \leqslant \left. \frac{d}{ds} \right|_{s=0} \left(\frac{|\gamma(s(P''(t)) - s, t) - \gamma(s(P_3(t)), t)|}{\ell - s} \right)$$

$$= \frac{1}{\ell} \left(\langle \boldsymbol{\omega}_2, \boldsymbol{e}_1 \rangle + \frac{d}{\ell} \right)$$

$$= \frac{1}{\ell} \left(\frac{d}{\ell} - 1 \right)$$

$$< 0,$$

which is impossible. Hence, $P''(t)$ must be interior.

As in Case (a), we want to show that the minimum of d/ℓ over $c_2(t) \times c_2(t)$ is attained in the interior whenever it is less than some small ε_0. Again, we assume the minimum of d/ℓ is attained on the boundary. The algebraic argument above shows that one may assume the line segment connecting $P_3(t)$ and $P''(t)$ has no interior intersection with $c_2(t)$. By minimality, we have

$$0 = \left. \frac{d}{ds} \right|_{s=0} \left(\frac{|\gamma(s(P''(t)) + s, t) - \gamma(s(P_3(t)), t)|}{\ell + s} \right)$$

$$= -\frac{1}{\ell} \left(\langle \boldsymbol{\omega}_2, \boldsymbol{e}'' \rangle + \frac{d}{\ell} \right) . \tag{8.31}$$

When the minimum is less than some suitably small ε_0, we see from (8.31) that $\boldsymbol{\omega}_2$ is nearly orthonormal to $\boldsymbol{e}''$. It is also clear that $P''(t)$ cannot lie on the subarc connecting $P_3(t)$ to $P_2(t)$. To avoid an interior intersection, $\boldsymbol{e}''$ must be almost horizontal, and so $\boldsymbol{\omega}_2$ is close to $-\boldsymbol{e}_3$. We compute

$$\frac{d}{ds}\Big|_{s=0}\left(\frac{|\gamma(s(P_3(t))+s,t)-\gamma(s(P''(t)),t)|}{\ell-s}\right)$$

$$=\frac{1}{\ell}\left(\langle\boldsymbol{\omega}_2,\boldsymbol{e}_3\rangle+\frac{d}{\ell}\right)$$

$$<\;0\,,$$

provided ε_0 is suitably small. So, d/ℓ has an interior minimum and this leads to a contradiction as before. We have finally finished the proof of the Theorem 8.1.   □

Notes

Whether the Grayson convexity theorem holds for the GCSF remains an open problem. In Angenent-Sapiro-Tannenbaum [19], it is proved that the affine CSF shrinks an embedded closed curve to a point with total absolute curvature converging to 2π. We have adapted and made straightforward generalizations of many results in this paper to $(8.1)_\sigma$, $\sigma\in(0,1)$. Another main ingredient in the proof of Theorem 8.1 is the monotonicity of Huisken's isoperimetric ratio [78]. We observe that it continues to hold for a large class of flows, including $(8.1)_\sigma$. See Remark 8.5.

Bibliography

This bibiography is not intended to be inclusive. For topics not directly related to the flows discussed in this book, we usually list a few representative papers or a recent work from which the reader can find more references.

[1] U. Abresch and J. Langer, The normalized curve shortening flow and homothetic solutions, J. Differential Geom. **23** (1986), 175–196.

[2] J. Ai, K.S. Chou and J. Wei, Self-similar solutions for the anisotropic affine curve shortening problem, to appear in Calc. Var. PDEs.

[3] S.J. Altschuler, Singularities of the curve shrinking flow for space curves, J. Differential Geom. **34** (1991), 491–514.

[4] S.J. Altschuler and M.A. Grayson, Shortening space curves and flow through singularities, J. Differential Geom. **35** (1992), 283–298.

[5] S.J. Altschuler and L.F. Wu, Convergence to translating solutions for a class of quasilinear parabolic boundary problems, Math. Ann. **295** (1993), 761–765.

[6] L. Alvarez, F. Guichard, P.L. Lions and J.M. Morel, Axioms and fundamental equation of image processing, Arch. Rational Mech. Ana. **132** (1993), 199–257.

[7] L. Ambrosio and H.M. Soner, Level set approach to mean curvature flow in arbitrary codimension, J. Differential Geom. **43** (1996), 693–737.

[8] B. Andrews, Contraction of convex hypersurfaces by their affine normal, J. Differential Geom. **43** (1996), 207–230.

[9] B. Andrews, Monotone quantities and unique limits for evolving convex hypersurfaces, Inter. Math. Res. Notices **20** (1997), 1001–1031.

[10] B. Andrews, Evolving convex curves, Calc. Var. PDEs **7** (1998), 315–371.

[11] B. Andrews, The affine curve-lengthening flow, J. Reine Angrew. Math. **506** (1999), 43–83.

[12] S.B. Angenent, The zero set of a solution of a parabolic equation, J. Reine Angrew. Math. **390** (1988), 79–96.

[13] S.B. Angenent, Parabolic equations for curves on surfaces, I: Curves with p-integrable curvature, Ann. of Math. **132** (2) (1990), 451–483.

[14] S.B. Angenent, Parabolic equations for curves on surfaces, II: Intersections, blow-up and generalized solutions, Ann. of Math. (2) **133** (1991), 171–215.

[15] S.B. Angenent, On the formation of singularities in the curve shortening flow, J. Differential Geom. **33** (1991), 601–633.

[16] S.B. Angenent, Inflection points, extatic points and curve shortening. **Hamiltonian systems with three or more degrees of freedom** (S'Agaró 1995), 3–10, NATO Adv. Sci. Inst. Ser. C Math. Phys. Sci. **553**, Kluwer, Dordrecht, 1999.

[17] S.B. Angenent and M.E. Gurtin, Multiphase thermomechanics with interfacial structure, 2. Evolution of an isothermal interface, Arch. Rational Mech. Anal. **108** (1989), 323–391.

[18] S.B. Angenent and J.J.L. Velazquez, Asymptotic shape of cusp singularities in curve shortening, Duke Math. J. **77** (1995), 71–110.

[19] S.B. Angenent, G. Sapiro and A. Tannenbaum, On the affine heat equation for non-convex curves, J. Amer. Math. Soc. **11** (1998), 601–634.

[20] W. Ballman, G. Thorbergsson and W. Ziller, Existence of closed geodesics on positively curved manifolds, J. Differential Geom. **18** (1983), 221–252.

[21] W. Blaschke, Über Affine Geometrie I: Isoperimetrische Eigenschaften von Ellipse and Ellipsoid. Ber. Verh. Sächs. Akad. Leipzig, Math. Phys. Kl. **68** (1916), 217–239.

[22] K.A. Brakke, **The motion of a surface by its mean curvature**, Princeton Univ. Press, Princeton, N.J., 1978.

[23] J.W. Cahn, C.M. Elliott and A. Novick-Cohen, The Cahn-Hilliard equation with a concentration dependent mobility: motion by minus the Laplacian of the mean curvature, European J. Appl. Math. **7** (1996), 287–301.

[24] J.W. Cahn and J.E. Taylor, Surface motion by surface diffusion, Acta Metallurgica **42** (1994), 1045-1063.

[25] E. Calabi, P.J. Olver and A. Tannenbaum, Affine geometry, Curve flows and invariant numerical approximations, Adv. Math. **124** (1996), 154–196.

[26] S.Y.A. Chang, M.J. Gursky and P.C. Yang, The scalar curvature equation on 2- and 3-spheres, Calc. Var. PDEs **1** (1993), 205–229.

[27] X. Chen, Generation and propagation of interfaces for reaction-diffusion equations, J. Differential Eqn's. **96** (1992), 116–141.

[28] Y.G. Chen, Y. Giga and S. Goto, Uniqueness and existence of viscosity solutions of generalized mean curvature flow equations, J. Differential Geom. **33** (1991), 749–786.

[29] K.S. Chou (K. Tso), Deforming a hypersurface by its Guass-Krouecker curvature, Comm. Pure Appl. Math. **38**, (1985), 867–882.

[30] K.S. Chou and Y.C. Kwong, On quasilinear parabolic equations which admit global solutions for initial data with unrestricted growth, to appear in Calc. Var. PDEs.

[31] K.S. Chou and G. Li, Optimal systems of group invariant solutions for the generalized curve shortening flows, to appear in Comm. Anal. Geom.

[32] K.S. Chou and X.J. Wang, A logarithmic Gauss curvature flow and the Minkowski problem, to appear in Analyse non linéaire.

[33] K.S. Chou and L. Zhang, On the uniqueness of stable ultimate shapes for the anisotropic curve shortening problem, Manu. Math. **102**(2000), 101–111.

[34] K.S. Chou and X.P. Zhu, Shortening complete plane curves, J. Differential Geom. **50** (1998), 471–504.

[35] K.S. Chou and X.P. Zhu, Anisotropic flows for convex plane curves, Duke Math. J. **97** (1999), 579–619.

[36] K.S. Chou and X.P. Zhu, A convexity theorem for a class of anisotropic flows of plane curves, Indiana Univ. Math. J. **48** (1999), 139–154.

[37] B. Chow, On Harnack's inequality and entropy for the Gaussian curvature flow, Comm. Pure Appl. Math. **44** (1991), 469–483.

[38] B. Chow and D.H. Tsai, Geometric expansion of convex plane curves, J. Differential Geom. **44** (1996), 312–330.

[39] C. Croke, Poincaré's problem and the length of the shortest closed geodesic on a convex hypersurface, J. Differential Geom. **17** (1982), 595–634.

[40] P. DeMottoni and M. Schatzman, Development of interfaces in $\mathbb{R}^N$, Proc. Roy. Soc. Edinburgh Sect. A **116** (1990), 207–220.

[41] K. Deckelnick, Weak solution of the curve shortening flow, Calc. Var. PDEs **5** (1997), 489–510.

[42] K. Deckelnick, C.M. Elliott and G. Richardson, Long time asymptotics for forced curvature flow with applications to the motion of a superconducting vortex, Nonlinearity **10** (1997), 655–678.

[43] E. DiBenedetto, **Degenerate Parabolic Equations**, Universitext series, Springer-Verlag, New York, (1993).

[44] C. Dohmen and Y. Giga, Self-similar shrinking curves for anisotropic curvature flow equations, Proc. Japan Aca. Ser. A **70** (1994), 252–255.

[45] C. Dohmen, Y. Giga and N. Mizoguchi, Existence of self-similar shrinking curves for anisotropic curvature flow equations, Calc. Var. PDEs **4** (1996), 103–119.

[46] K. Ecker and G. Huisken, Mean curvature evolution of entire graphs, Ann. of Math. (2) **130** (1989), 453–471.

[47] K. Ecker and G. Huisken, Interior estimates for hypersurfaces moving by mean curvature, Invent. Math. **105** (1991), 547–569.

[48] C.L. Epstein and M. Gage, The curve shortening flow, **Wave Motion : Theory, Modeling and Computation**, A Chorin and A Majda, Editors, Springer-Verlag, New York, 1987.

[49] C.L. Epstein and M.I. Weinstein, A stable manifold theorem for the curve shortening equation, Comm. Pure Appl. Math. **40** (1987), 119–139.

[50] L.C. Evans and J. Spruck, Motion of level sets by mean curvature I, J. Differential Geom. **33** (1991), 635–681.

[51] W.J. Firey, Shapes of worn stones, Mathematica **21** (1974), 1–11.

[52] M.L. Frankel and G.I. Sivashinsky, On the equation of a curved flame front, Phys. D **30** (1988), 28–42.

[53] A. Friedman, **Partial Differential equations of Parabolic Type**, Krieger, Malabar, FL, 1983.

[54] M.E. Gage, An isoperimetric inequality with applications to curve shortening, Duke Math. J. **50** (1983), 1225–1229.

[55] M.E. Gage, On an area-preserving evolution equation for plane curves, **Nonlinear problems in geometry** (Mobile, Ala., 1985), 51–62, Contemp. Math. 51, Amer. Math. Soc., Providence, R.L., 1986.

[56] M.E. Gage, Deforming curves on convex surfaces to simple closed geodesics, Indiana Univ. Math. J. **39** (1990), 1037–1059.

[57] M.E. Gage, Evolving plane cuves by curvature in relative geometries, Duke Math. J. **72** (1993), 441–466.

[58] M.E. Gage and R.S. Hamilton, The heat equation shrinking convex plane curves, J. Differential Geom. **23** (1986), 69–96.

[59] M.E. Gage and Y. Li, Evolving plane curves by curvature in relative geometries II, Duke Math. J. **75** (1994), 79-98.

[60] C. Gerhardt, Flow of nonconvex hypersurfaces into spheres, J. Differential Geom. **32** (1990), 299–314.

[61] M.H. Giga and Y. Giga, Evolving graphs by singular weighted curvature, Arch. Rational Mech. Anal. **141** (1998), 117–198.

[62] M.H. Giga and Y. Giga, Crystalline and level set flow–Convergence of a crystalline algorithm for a general anisotropic curvature flow in the plane, Gakuto International Series, Mathematical Sciences and Applications Vol. 13 (2000), **Free Boundary Problems: Theory and Applications**, 64–79. (ed. N. Kenmochi), Proc. of FBP'99.

[63] Y. Giga, S. Goto, H. Ishii and M.H. Sato, Comparison principle and convexity preserving properties for singular degenerate parabolic equations unbounded domains, Indiana Univ. Math. J. **40** (1991), 443–470.

[64] Y. Giga and K. Ito, Loss of convexity of simple closed curves moved by surface diffusion, Topics in nonlinear analysis, 305–320. **Progr. Nonlinear Differential Equations Appl. 35**, Birkhauser, Basel 1999.

[65] Y. Giga and R.V. Kohn, Asymptotically self-similar blow-up of semilinear heat equations, Comm. Pure Appl. Math. **38** (1985), 297–319.

[66] M.A. Grayson, The heat equation shrinks embedded plane curves to round points, J. Differential Geom. **26** (1987), 285–314.

[67] M.A. Grayson, The shape of a figure-eight under the curve shortening flow, Invent. Math. **96** (1989), 177–180.

[68] M.A. Grayson, Shortening embedded curves, Ann. of Math. **129** (1989), 71–111.

[69] R.E. Goldstein and D.M. Petrich, The Korteweg-de Vries hierarchy as dynamics of closed curves in the plane, Phys. Rev. Lett. **67** (1991), 3203–3206.

[70] H.W. Guggenheimer, **Differential Geometry**, McGraw-Hill, New York, 1963.

[71] M.E. Gurtin, **Thermomechanics of evolving phase boundaries in the plane**, Oxford Mathematical Mono., Clarendon, Oxford, 1993.

[72] R.S. Hamilton, Isoperimetric estimates for the curve shrinking flow in the plane, **Modern Methods in Complex Analysis** (Princeton, 1992), 201–222, Ann. of Math. Stud. 137, Princeton Univ. Press, Princeton, N.J. 1995.

[73] R.S. Hamilton, The formation of singularities in the Ricci flow, **Surveys in differential geometry** Vol. II (Cambridge, MA, 1993), 7–136, Internat. Press, Cambridge, MA, 1995.

[74] H. Hasimoto, A soliton on a vortex filament, J. Fluid Mech. **51** (1972), 477.

[75] N. Hungerbuhler and K. Smoczyk, Soliton solutions for the mean curvature flow, preprint 1998.

[76] G. Huisken, Non-parametric mean curvature evolution with boundary conditions, J. Diff. Eqn's. **77** (1988), 369–378.

[77] G. Huisken, Asymptotic behaviour for singularities of the mean curvature flow, J. Differential Geom. **31** (1990), 285–299.

[78] G. Huisken, A distance comparison principle for evolving curves, Asian J. Math. **2** (1998), 127–133.

[79] G. Huisken and C. Sinestrari, Convexity estimates for mean curvature flow and singularities of mean convex surfaces, Acta Math. **183** (1999), 45–70.

[80] R. Ikota, N. Ishimura and T. Yamaguchi, On the structure of steady solutions for the kinematic model of spiral waves in excitable media, Japan J. Indust. Appl. Math. **15** (1998), 317-330.

[81] T. Ilmanen, Elliptic regularization and partial regularity for motion by mean curvature, Mem. Amer. Math. Soc. **108** (1994), no. 520.

[82] J.P. Keener, Symmetric spirals in media with relaxation kinetics and two diffusion species, Phys. D. **70** (1994), 61–73.

[83] J.P. Keener and J.J. Tyson, The dynamics of scroll waves in excitable media, SIAM Rev. **34** (1992), 1–39.

[84] J.P. Keener and J. Sneyd, **Mathematical Physiology**, Springer-Verlag, NY 1998.

[85] W. Klingenberg, **Lectures on Closed Geodesics**, Grundlehren Math. Wiss. 230, Springer-Verlag, Berlin 1978.

[86] O.A. Ladyzhenskaya, V.A. Solonnikov and N.N. Ural'ceva, **Linear and Quasilinear Equations of Parabolic Type**, Amer. Math. Soc. Providence, RI, 1968.

[87] J. Langer and D.A. Singer, Curve Straightening and a minimax argument for closed elastic curves, Topology **24** (1985), 241–247.

[88] G.M. Lieberman, **Second Order Parabolic Differential Equations**, World Scientific, Singapore 1996.

[89] E. Meron, Pattern formation in excitable media, Phys. Rep. **218** (1992), 1–66.

[90] J.D. Murray, **Mathematical Biology**, Springer-Verlag, NY, 1989.

[91] K. Nakayama, H. Segur and M. Wadati, Integrability and the motion of curves, Phys. Rev. Lett. **69**, 2603–2606.

[92] J.A. Oaks, Singularities and self-intersections of curves evolving on surfaces, Indiana Univ. Math. J. **43** (1994), 959–981.

[93] P. J. Olver, **Equivalence, Invariants and Symmetry**, Cambridge Univ. Press, Cambridge, 1995.

[94] P.J. Olver, G. Sapiro and A. Tannenbaum, Differential invariant signatures and flows in computer vision: A symmetry group approach, **Geometry-Driven Diffusion in Computer Vision**, B.M. Ter Haar Romeny, Ed., 205–306, Kluwer, Dordrecht, 1994.

[95] A. Polden, Evolving curves, Honours thesis, Australian National Univ. 1991.

[96] M.H. Protter and H.F. Weinberger, **Maximum Principles in Differential Equations**, Prentice-Hall, 1967.

[97] G. Sapiro and A. Tannenbaum, On invariant curve evolution and image analysis, Indiana Univ. J. Math. **42** (1993), 985–1009.

[98] G. Sapiro and A. Tannenbaum, On affine plane curve evolution, J. Funct. Anal. **119** (1994), 79–120.

[99] R. Schneider, **Convex Bodies: The Brunn-Minkowski Theory**, Cambridge Univ. Press, Cambridge, 1993.

[100] H.M. Soner, Motion of a set by the curvature of its boundary, J. Diff. Eqn's **101** (1993), 313–372.

[101] P.E. Souganidis, Front propagation: Theory and applications, 186–242, Lecture Notes in Math. 1660, Springer-Verlag, Berlin, 1997.

[102] A. Stahl, Convergence of solution to the mean curvature flows with a Neumann boundary condition, Calc. Var PDEs **4** (1996), 421–441.

[103] A Stancu, Uniqueness of self-similar solutions for a crystalline flow, Indiana Univ. Math. J. **45** (1996), 1157–1174.

[104] A. Stancu, Asymptotic behaviour of solutions to a crystalline flow, Hokkaido Math. J. **27** (1998), 303–320.

[105] M. Struwe, On the evolution of harmonic maps in higher dimensions, J. Differential Geom. **28** (1988), 485–502.

[106] J.E. Taylor, Constructions and conjectures in crystalline nondifferential geometry, **Differential Geometry**, 321–336, Pitman Mono. Surveys Pure Appl. Math., **52**, Longman, Harlow, 1991.

[107] J.E. Taylor, Motion of curves by crystalline curvature, including triple junctions and boundary points, **Diff. Geom.: Partial Diff. Eqn's on Manifold** (Los Angeles, 1990), 417–438, Proc. Sympos. Pure Math. **54**, Part 1, Amer. Math. Soc., Providence, RI, 1993.

[108] J.I.E. Urbas, On the expansion of starshaped hypersurfaces by symmetric function of their principal curvatures, Math. Z. **205** (1990), 355–372.

[109] J.I.E. Urbas, Convex curves moving homothetically by negative powers of their curvature, Asian J. Math. **3** (1999), 635–658.

[110] Y. Wen, L^2-flow of curve straightening in the plane, Duke Math. J. **70** (1993), 683–698.

[111] Y. Wen, Curve straightening flow deforms closed plane curves with non-zero rotation number to circles, J. Differential Eqn's. **120** (1995), 89–107.

[112] E. Zeilder, **Nonlinear Functional Analysis and its Applications**, Vol. 1, Springer-Verlag, NY, 1985.

[113] X.P. Zhu, Aysmptotic behaviour of anisotropic curve flows, J. Differential Geom. **48** (1998), 225–274.

Index

9 780367 397531